Advances in 21st Century Human Settlements

Indexed by SCOPUS

This Series focuses on the entire spectrum of human settlements – from rural to urban, in different regions of the world, with questions such as: What factors cause and guide the process of change in human settlements from rural to urban in character, from hamlets and villages to towns, cities and megacities? Is this process different across time and space, how and why? Is there a future for rural life? Is it possible or not to have industrial development in rural settlements, and how? Why does 'urban shrinkage' occur? Are the rural areas urbanizing or is that urban areas are undergoing 'ruralisation' (in form of underserviced slums)? What are the challenges faced by 'mega urban regions', and how they can be/are being addressed? What drives economic dynamism in human settlements? Is the urban-based economic growth paradigm the only answer to the quest for sustainable development, or is there an urgent need to balance between economic growth on one hand and ecosystem restoration and conservation on the other – for the future sustainability of human habitats? How and what new technology is helping to achieve sustainable development in human settlements? What sort of changes in the current planning, management and governance of human settlements are needed to face the changing environment including the climate and increasing disaster risks? What is the uniqueness of the new 'socio-cultural spaces' that emerge in human settlements, and how they change over time? As rural settlements become urban, are the new 'urban spaces' resulting in the loss of rural life and 'socio-cultural spaces'? What is leading the preservation of rural 'socio-cultural spaces' within the urbanizing world, and how? What is the emerging nature of the rural-urban interface, and what factors influence it? What are the emerging perspectives that help understand the human-environment-culture complex through the study of human settlements and the related ecosystems, and how do they transform our understanding of cultural landscapes and 'waterscapes' in the 21st Century? What else is and/or likely to be new vis-à-vis human settlements – now and in the future? The Series, therefore, welcomes contributions with fresh cognitive perspectives to understand the new and emerging realities of the 21st Century human settlements. Such perspectives will include a multidisciplinary analysis, constituting of the demographic, spatio-economic, environmental, technological, and planning, management and governance lenses.

If you are interested in submitting a proposal for this series, please contact the Series Editor, or the Publishing Editor:

Bharat Dahiya (bharatdahiya@gmail.com) or
Loyola D'Silva (loyola.dsilva@springer.com)

More information about this series at http://www.springer.com/series/13196

Innocent Chirisa · Andrew Chigudu
Editors

Resilience and Sustainability in Urban Africa

Context, Facets and Alternatives in Zimbabwe

Editors
Innocent Chirisa
Department of Demography Settlement
and Development
University of Zimbabwe
Mount Pleasant, Zimbabwe

Andrew Chigudu
Department of Architecture and Real Estate
University of Zimbabwe, Harare,
Zimbabwe

ISSN 2198-2546 ISSN 2198-2554 (electronic)
Advances in 21st Century Human Settlements
ISBN 978-981-16-3290-7 ISBN 978-981-16-3288-4 (eBook)
https://doi.org/10.1007/978-981-16-3288-4

This Springer imprint is published by the registered company Springer Nature Singapore Pte Ltd.
The registered company address is: 152 Beach Road, #21-01/04 Gateway East, Singapore 189721, Singapore

Contents

Editors and Contributors

About the Editors

Innocent Chirisa is a full professor at the Department of Rural and Urban Planning, University of Zimbabwe. He is currently the acting dean of the Faculty of Social Studies at the University of Zimbabwe and a research fellow at the Department of Urban and Regional Planning, University of the Free State, South Africa. His research interests are systems dynamics in urban land, regional stewardship and resilience in human habitats.

Andrew Chigudu is a D.Phil. student and lecturer with the University of Zimbabwe. He is an academic and seasoned spatial planning practitioner. He is the executive director for Full Life Open Arms Africa Investment since November 2019. Before this, he was the managing director for Hello Project Developers for eight years. He has authored and co-authored in refereed journals. He is the current vice-president for Zimbabwe Institute for Regional and Urban Planning (ZIRUP).

Contributors

Elmond Bandauko Department of Geography and Environment, The University of Western Ontario, London, ON, Canada;
Center for Urban Policy and Local Governance, The University of Western Ontario, London, Ontario, Canada

Tinashe Bobo Harare, Zimbabwe

Andrew Chigudu Department of Architecture and Real Estate, University of Zimbabwe, Arlington, Harare, Zimbabwe

Witness Chikoko Department of Social Work, University of Zimbabwe, Harare, Zimbabwe

Godfrey Chikowore Faculty of Arts, Midlands State University, Zvishavane Campus, Zimbabwe

Innocent Chirisa Department of Demography Settlement and Development, University of Zimbabwe, Mt Pleasant, Harare, Zimbabwe;
Faculty of Natural and Agricultural Sciences, Bloemfontein, Republic of South Africa

Christine R. Chivandire Department of Demography Settlement and Development, University of Zimbabwe, Mt Pleasant, Harare, Zimbabwe

Audrey N. Kwangwama Harare, Zimbabwe

Shamiso Hazel Mafuku Department of Architecture and Real Estate, University of Zimbabwe, Harare, Zimbabwe

Pranitha Maharaj School of Built Environment and Development Studies, University of KwaZulu-Natal, Durban, South Africa

Emma Maphosa Department of Architecture and Real Estate, University of Zimbabwe, Mt Pleasant, Harare, Zimbabwe

Blessing Marandure School of Applied Social Science, Faculty of Health and Life Science, De Montfort University, Leicester, UK

Terence Motida Mashonganyika Faculty of Arts, Midlands State University, Zvishavane Campus, Zimbabwe

Thebeth R. Masunda Department of Community and Social Development, University of Zimbabwe, Harare, Zimbabwe

Abraham Rajab Matamanda Department of Geography, Bloemfontein, Republic of South Africa

Washington M. Mbaura Department of Demography Settlement and Development, University of Zimbabwe, Harare, Zimbabwe

Samson Mhizha Department of Applied Pyschology, University of Zimbabwe, Harare, Zimbabwe

Tinashe N. Mujongonde-Kanonhuhwa Department of Demography Settlement & Development, University of Zimbabwe, Harare, Zimbabwe

Patience Mukuzunga Department of Demography Settlement and Development, University of Zimbabwe, Mt Pleasant, Harare, Zimbabwe

Yvonne Munanga Department of Architecture and Real Estate, University of Zimbabwe, Harare, Zimbabwe

Langton Mundau Department of Social Work, University of Zimbabwe, Harare, Zimbabwe

Constantine Munhande Faculty of Arts, Midlands State University, Zvishavane Campus, Zimbabwe

Roselyn Ncube Women's University in Africa, Harare, Zimbabwe

Verna Nel Faculty Natural and Agricultural Sciences, Bloemfontein, Republic of South Africa

John David Nhavira Faculty of Business Management Sciences and Economics, University of Zimbabwe, Harare, Zimbabwe

Chapter 1
Resilience and Sustainability in Urban Land Dynamics in Africa: A Review

Innocent Chirisa and Verna Nel

Abstract This chapter puts, at the centre, dynamics in urban land as a critical factor in determining resilience and sustainability with reference to Africa. It traces the history and dynamics of land markets in Zimbabwe with consideration to resilience and sustainability. The chapter argues that what is known is that, Zimbabwean cities are heterogeneous and shaped by multiple factors including urbanisation, macroeconomic and political factors. Results reveal that the practices fail to ensure a sustainable urban land markets due to a host of challenges ranging from insecure tenure, failure to adapt to social and economic environment dynamics, lack of political will to institutional challenges where policy that influence land-use changes is often driven by the interests of the elites at the expense of the poor. To develop and maintain effective land markets, several critical factors have to be in place. The land registration system must also be affordable so that all citizens, especially women and minority groups, rich and poor, can have access to it. Continuous globalisation will inescapably shock land markets, especially as information technology, including web-based services, provides greater access to national land information services.

Introduction

Land is distinct from other properties that can be bought or sold because it is immoveable. Buildings upon land can be dismantled and moved elsewhere but the land on that they stand cannot be moved or destroyed. Only rights to use land can be traded through land markets. Urban land markets exist when and wherever it is possible to exchange land rights for agreed amounts of money or any form of equivalent payment (Mahoney et al. 2007). Since land is immovable, urban land markets are prone to

I. Chirisa (✉)
Department of Demography Settlement & Development, University of Zimbabwe, Mt Pleasant, PO Box MP167, Harare, Zimbabwe

I. Chirisa · V. Nel
Faculty of Natural and Agricultural Sciences, PO Box, Bloemfontein 9300, Republic of South Africa

I. Chirisa and A. Chigudu (eds.), *Resilience and Sustainability in Urban Africa*, Advances in 21st Century Human Settlements,
https://doi.org/10.1007/978-981-16-3288-4_1

in-situ factors. The efficiency of urban land markets varies across the globe, on the openness of the land market to public inquiry and its support for sustainable development. The increase in informal settlements in African cities has made it difficult to develop effective land markets. The rise in informal settlements is due to the fact urban planning practices in many African cities rarely reflect the realities of urban Africa as they fail to take into account the social, political, economic, and environmental milieu of the continent's urban development [World Economic Forum (WEF, 2017)].

The volume of urban planning and building codes reflect contradictory, complex and outmoded colonial planning standards. Furthermore, customary practice, unregulated regimes, weak capacity of local authorities and poor strategic focus have led to cities being built from the "back to front" because construction occurs prior to urban planning. Poor planning has also posed serious challenges for sustainable city development and contributes to lofty costs of real estate and housing in African cities. Consequently, the quality of life for citizens deteriorates whilst the competitiveness of firms and their workforce is negatively affected. Subsequently, much of Africa's urban expansion is occurring through spontaneous transformation of rural land into urban land. This type of transformation affects urban land markets, urban resilience and sustainability (Landman & Nel, 2012; Lawrence, 2014; Klein et al., 2017).

The Zimbabwean case shows no difference. There is uncertainty regarding the scope of urbanisation and urban growth because of gaps in knowledge and data. What is known is that, Zimbabwean cities are heterogeneous and shaped by multiple factors, such as the nature of urbanisation, macroeconomic and political factors, including the sturdy influence of informality (DFID, 2017). These factors shape urban land markets. This chapter, therefore, traces the history and dynamics of land markets in Zimbabwe with consideration to resilience and sustainability.

Resilience has become a very topical issue transcending many spheres and sectors of development. It is applicability to land, environment, spatial modelling and planning has grown tremendously (White & Engelen, 2000; Williamson, 2001; Wilkinson, 2012a, 2012b; White & O'Hare, 2014). If urban land is regarded as a system, then the relevance of resilience analysis is quite central (Nijkamp et al., 2002; Nel and Nel, 2012; Silva & Wu, 2012).

In informal settlements, the relevance of resilience is acknowledged and should be instrumental for sustainable land management practices (Woolf et al., 2016; World Bank, 2012, 2013). Resilience is considered a critical urban design matter (Caputo et al., 2015; O'Hare & White, 2013; Shaw, 2012; Stumpp, 2013; Wu & Wu, 2013). Activities, like housing development and extensions, transform urban landscapes and are instrumental in the creation of land values (or deterioration thereof) (Tipple, 2000). In this place-making endeavour, ideology of the government of the day matters, for example, the days of apartheid in South Africa and thereafter (Turok, 2001; UN-Habitat, 2010). The apartheid ideology was all about separate development according to race. The 'capitalist' ideology after South Africa achieved democracy in 1994 has now become a matter of separate development by income. The poor live in slums and townships while the rich reside in gated-communities and posh and leafy suburbs. Resilience is often thought of when one is dealing with natural risks and disasters (Van Niekerk, 2013; Walker & Salt, 2006; Walker et al., 2006), yet policy-led disasters that impinge on the markets could also be considered as part of the equation.

Theoretical Framework

Land markets are shaped by land tenure and land administration systems. Key fundamentals in any land market system include land turing factors, such as the extent to which land can be bought or sold and consideration of whether land can be used as collateral and if so, who takes over a property in case of a loan default. They are also issues about rights of servitude and easements, rules governing inheritance, and the extent of additional use rights and obligations (Van der Molen, 2006). The viability of a land market depends on the number of transactions per time as.

Other than land tenure, land administration systems are also important in shaping urban land markets. Besides providing security of tenure, stable and proper land administration is necessary in building certainty and making land markets dependable and consequently trustworthy, that is an essential component for stimulating investment and transactions in land markets. Flaws in land administration can have extensive consequences (Williamson, 2001; Van der Molen, 2006; Bennett et al., 2008). Urban land markets are not static but, evolve due to changes in certain factors, such as population growth, economic development, social and political dynamics and land management policies (Xie & Sun, 1999: 208–209). An in-depth understanding of such a system is therefore necessary since, in the absence of such an understanding of land markets, it is difficult to identify processes involved in the determination, recording and diffusion of information about tenure arrangements. Yet information on planning services is needed to guarantee security of tenure, markets, planning, taxation and management of resources.

Literature Review

Resilience is a proactive approach to strengthen the ability to resist, absorb, and recover from shocks or hazards without jeopardising socio-economic development (Abhas, Miner and Stanton-Geddes, 2013; Seeliger & Turok, 2013, 2014; Sellberg et al., 2015). Sustainability is the capacity of responsible authority to operate, maintain, renew or expand its service delivery system for the vulnerable (Omwamba et al., 2015). Urban resilience and sustainability therefore, are issues of concern given that urban areas are focal points, drivers of change and engines of economic growth (Martin-Breen & Anderies, 2011; Maclean et al., 2014; Mcphearson et al,. 2015; Meerow et al., 2016). They generate more than 50% of most countries' Gross Domestic Product (UN-Habitat, 2015). With the global population living in cities having crossed the 50% mark in 2008, present day cities have to address the needs of millions (Brown & Dixon, 2014). Nation states have vested interests on the well-being of cities since they are the major source of national capital (Campanella, 2006).

Regardless of the focus on city wellness by nation states, a city's resilience is affected by a number of factors that vary from place to place (Campanella, 2006). Some of these factors include nature of property ownership, strength of the insurance industry, urban planning and design and the socio-political and economic spectrum. Some of the issues influencing resilience of property markets are products of macro-political and economic realities that cannot be easily changed (Cote & Nightingale, 2012; Coutu, 2002; Davoudi et al., 2012; De Weijer, 2013; Du Plessis, 2008). A typical example is the privatisation of urban property markets that results in a more lucid documentation of property. Fee-simple ownership of private property created an indestructible means of organising space, through legal credentials, rather than physical ownership (Campanella, 2006:143). These developments in land tenure coupled with a stronger insurance industry, makes reconstruction quicker, hence improves urban resilience. The other positive impact of insurance is that it urges property owners to rebuild on brownfields since insurance awards are based on what was lost and where (*ibid.*). At the same time, geographic and economic advantages that prompted a city's initial development also works in favour of promoting a city's resilience. A city with a robust and diversified economy rebounds quicker than oney with a narrowly specialised, weak economy.

Apart from the economy, architectural design quality starting from site to city level also influences resilience. Usually, concrete foundations and deeply buried utilities frequently survive major catastrophes and ensure that cities are resilient. Cities investing in environmental risk management planning reduce vulnerability of both physical infrastructure assets and human life (Campanella, 2006). However, the resilience of a city is not only limited to the planning and design of buildings but also stretches to the socio-cultural character of a city (Campanella, 2006). This is because cultural values and institutions shape the physical form of a city. Though it is one thing for a city's buildings to be destroyed, it is worse for its cultural institutions and social fabric to be dismantled. To permit total recovery, a connection should be made between values across family, society and religious systems (Du Plessis, 2012; Du Plessis & Brandon, 2015; Du Plessis & Cole, 2011; Elmqvist, 2014; Ernstson et al., 2010).

Urban Land Market Dynamics

Activities in urban systems follow life cycles. This applies to both the physical stock and activities or uses of properties (Batty et al., 1999). In the urban dynamics model, activities can be classified as new, established and declining housing, industry and commercial land uses. Each use is subject to different rates of growth (or decline) and different scales by that existing activities are attracted or detracted from new growth in related locations" (Forrester, 1969). Systems in urban areas are therefore not static but change due to different stimuli.

One example of classification of a city based on activity is an entrepreneurial city. An entrepreneurial city is sensitive to the needs of the business sector, has a sturdy sense for innovativeness and elasticity, is project-oriented, strategic in temperament, seeks co-operation with the private sector, aims to create its added value and monitors its own socio-economic performance (Nijkamp et al., 2002). Given the multifunctional and varied nature of urban economies, a solitary and unambiguous performance indicator is hard to identify. It is undeniable that cities have always been powerhouses of economic activity and a source of innovation (Nijkamp et al., 2002: 1865). The urban economics literature shows that agglomerative economies offer locational advantages for business entrepreneurs. Furthermore, the 'new economic geography' places urban economics at the centre of international trade and networks. The emergence of new economic opportunities for cities has encouraged a world-wide debate on the 'new mission' of urban governments in our modern twenty-first century.

Repositioning urban policy seems to be fundamental, with the interface between the public and the private sectors being very critical. The minimal model of a city demands two layers, one for the city's infrastructure units, and the other for migrating populations. Urban infrastructure (land units, network segments, buildings) are immobile while humans are free to change their locality in the city (Benenson, 1998:25). From the economic point of view, the city evolves towards distributing individuals in segregated locations based on economic capabilities (Benenson, 1998: 34). At one extreme end, is a group of agents characterised by low economic status, minimal rate of status growth, and occupying houses of low value. At the other extreme end is a group of high-status agents, different values, optimal growth rate. These occupy neighbourhoods with high-value properties. The distance between the neighbourhoods occupied by high- and low-status agents is proportionate to the economic differences between these groups.

The question of how to re-use the spaces of confinement in cities often boils down to slightly more than the question of land rent (Bond, 2000). It leads to other questions on how urban capitalism works. Land rent reflects the return on capital that can be accredited to locational and spatial characteristics of a place. Nevertheless, capital is by nature movable, a condition deemed necessary for its reproduction over time. Movability of capital is also a condition necessary to preserve the class power of the bourgeoisie against that of labour driving "uneven development" (Smith, 1990:2).

Rent gradients measure influence of land markets on urban form and urban land market efficiency. "Land rent gradients is expressed as $P_X = P_0 e^{-d}$ where, P_x is the land price per metre at a distance of x to the city centre; P_0 is the land price per metre at the city centre; and d is the distance at x to the city centre" (Ding, 2004: 1898). "Land-use modelling reproduces the standard result that when homogeneous traders locate at the same place, transaction prices (land rents) are the same at locations equidistant from the CBD and land rents decline monotonically as distance from the CBD increases" (Filatova et al., 2009:17). The land rent gradient is estimated through regression analysis, where generated transaction prices are the dependent variable while distance from the CBD is the independent variable. The extent of the urban area is determined by the location where the bid of the highest-value buyer is just equal to the willingness-to-accept of the seller (the opportunity cost of land in a non-urban use).

The Gentrification and Downward Raiding Problem

Downward raiding is defined by "displacement and exclusion [as] middle-income groups 'raid' lower-income areas (often state-subsidised or informal settlements) and undertake service/infrastructural upgrades, low-income residents are both displaced and excluded" (Lemanski, 2014: 2946). Although downward raiding of state-subsidised settlements can positively present opportunity for incorporation of properties into the formal and high value property market zones and upward mobility of low-income groups, it can also make it difficult for the poor to secure houses in areas originally intended for them. As higher income purchasers improve and resell property, they successfully exclude low-income beneficiaries in the short- and long-term (Lemanski, 2014).

Three Generations of Urban Renewal Policies

There are three generations of urban renewal policy. These include the first generation (the era of the bulldozer), the second generation (neighbourhood rehabilitation) and third generation (revitalisation). These generations were a succession of the other but applied distinct approaches.

The first generation is the era of the bulldozer (Carmon, 1999). The major characteristics of policy in this generation is an emphasis on the built environment. Poor housing conditions and old buildings in the growing cities led to the need to make "better use" of central urban land by forcing the poor out of cities. This gave birth to slum clearance. In the United Kingdom, the process started on a massive scale following the Greenwood Act of 1930. In the US, there are contrary views on emergency of the first-generation urban renewal policy. Whilst some attribute the first generation of property development to the Housing Law of 1937, others attribute it to the 1949 legislation that gave the government responsibility for provision of decent and affordable housing to citizens. Consequently, over a quarter of a million housing units were demolished or sealed up and more than one and a quarter million people were rehoused in the UK in the 1930s (Gibson & Langstaf, 1982). The momentum was halted by the Second World War, to be renewed with the Housing Law of 1954.

The objective established by the planners, at that time. was to raze between 12 000 and 60,000 units a year and build between 100,000 and 150,000 new units (Gibson & Langstaf, 1982) Most of the demolished units were low-rise private property, while most of the new flats were in big blocks of public housing (Short, 1982). The Planning Policy Guidance Note 3 passed in 2000 in UK forced local authorities to engage in construction of approximately 60% housing within their brownfields

(Jones & Evans, 2008). Public authorities in the UK managed both the clearance and provision of housing for those relocated in new council housing. In the US, by contrast, concentration and clearance of land sites was done by public agencies, while the new construction was in the hands of private entrepreneurs. Resultantly, the number of apartments demolished under urban renewal programmes in the US were much greater than the number of units built. Slum areas were frequently replaced by shopping centres, public buildings, and cultural and entertainment centres, that were all in high demand in the booming years that followed World War II (Gibson & Langstaf, 1982). The few housing developments constructed were designated for people within a higher socio-economic class than those who were relocated (Greer, 1965). Between 1949 and 1964, only 0,5% expenditures by the American federal government for urban renewal was used for the relocation of the families and individuals displaced from their homes (Gans, 1967).

Despite the considerable differences in the nature of activities in the US and UK, the disparagement articulated against most of the projects was analogous (Wilmott & Young, 1957). The executors were criticised for ignoring the heavy psychological cost of imposed relocation and the social cost of the obliteration of healthy communities (Gans, 1962; Fried, 1966; Hartman, 1971; Parker, 1973; English, 1976; Hyra, 2012). In those cases where new residential neighbourhoods were built, planners and designers were blamed for building human multi-storey blocks that were unfit for family life, and unquestionably not appropriate for poor families. Moreover, in many areas, redevelopment projects continued for 2–3 decades and during that period, unused buildings and vacant lots covered the city centre causing vast economic and social damages. Similar disparagement for the construction of roads and commercial buildings in place of housing was observed in Canada, where the urban renewal plan included 48 projects between 1948 and 1968 (Carter, 1991). In France, the "modernisation" approach informing urban renewal activities between 1958–1975 resulted in exclusion (Primus & Metselaar, 1992). In many of the Western world's large cities, especially in the US, luxurious projects of concrete, steel and glass were built on the sites of slums razed by the bulldozers (Sanders, 1980). Thus, in many of the case studies, long-term costs (social and economic) of dislocation and concentration of poor people in large residential blocks were high. Therefore, the bulldozer approach as a leading regeneration strategy was condemned and disqualified in many of the places it was applied.

Failure of the bulldozer approach gave way to the second-generation policies. The second-generation movement focused on neighbourhood rehabilitation that is an inclusive approach emphasising orientation towards social problems, rather than the built form. Since the 1960s, a new approach to assisting distressed neighbourhoods was developed and implemented in the United States and in other countries (Cullingworth, 1973). It was an alternative to the failure of the First Generation. Its core values included economic growth, upward mobility of society and 'rediscovery of poverty' within the 'society of plenty' (Harrington, 1962).

Public opinion towards programmes that required large allocations for social welfare purposes became more favourable than those under the first generation. Consequently, inclusive rehabilitation programmes, intended to improve existing poor housing and urban environments were planned and implemented. In addition to housing rehabilitation, attention was also given to address social predicaments of the population by adding improving quality of social services. Though many of the projects successfully made public participation a leading agenda in policy-making, the great society programmes of the American President, Lyndon Johnson, under the banner of 'War on Poverty', did not succeed in preventing the riots that broke out in the mid-1960s in the large cities of the US. The response of the administration was the introduction of the Model Cities programmes (Haar, 1975).

This programme, that was funded by the Federal Government (80%) and local authorities (20%), established a comprehensive approach to the problems of poverty in the distressed areas (Carmon, 1999). The social programmes were influenced by ideas developed by American planners, such as participation of residents in community development. In 1975, there were 3750 projects related to WPoverty, with a combined budget that reached 34 million pounds. Various ministries of the British government and local authorities took part, dealing with matters of education, employment and welfare within the framework of the Urban Programme (Gibson & Langstaf, 1982). Most of the programmes were local in scale. Many of the projects were implemented in areas where the General Improvement Areas (GIA)—the main governmental programme for housing and environmental improvement in troubled areas) was already active. Consequently, housing renovation programmes were implemented simultaneously with others on social issues. This was a rare combination in Britain, where physical and social programmes were separated administratively and spatially.

Within seven years, 2.3 billion dollars was spent on targeted neighbourhoods under the management of the Department of Housing and Urban Development (HUD) (Frieden & Kaplan, 1975). Funding was mainly allocated to social projects in education, health, professional training and public safety. Only a small portion of the funds was spent on housing and infrastructure rehabilitation (Listokin, 1983). Despite the large quantity of good will and the large sums spent t, the public and governments were disappointed by the results of the research that were unable to indicate significant positive results for many of the large social programmes of the 1960s. One example of such a research is the work of Gibson and Prathes (1977), that assessed many evaluations of social programmes and concluded that 'nothing works'. Another is Charles Murray's conclusion that the only thing War on Poverty programmes managed to produce was more poor people (Murray, 1984). Failure of the programme was due to expansion of the framework from a model programme of 36 neighbourhoods to 66 and, later on, to double the number of neighbourhoods, almost without additional resources (Banfield, 1974). The other reason was that the programme, like others, was limited by its theoretical regulations (Moynihan, 1969). Besides these short-term failures, the programme had positive long-term consequences (Wood, 1990). Overall, the lacuna amid promise and performance was evidently large (Kaplan & Cuciti, 1986; Frieden & Kaplan, 1975: 234).

Owing to this, right-wing governments cancelled Second-Generation type programmes, and only slight public attention was paid to the worsening urban problems, especially in the core of cities. Between the 1970s and 1980s, interesting impulsive processes of revitalization occurred in large cities of the developed countries. In the UK, similar socio-economic forces were active during the 1960s and 1970s, creating similar responses in urban regeneration. In the physical realm, the salient trend was a transition from clearance to renovation of existing buildings and environments (Murrie, 1990). Renovations took place under the slogan 'old houses into new homes'. Many of the upgrading programmes in European countries were uni-sectorial and exclusively paid attention to physical renovation of housing and infrastructure (Alterman, 1991). This was experienced in Sweden, Holland and West Germany, with a few rare exceptions (Schmoll, 1991). In other countries, including Canada, France and Israel, the comprehensive model of the United States was applied. Canada's Neighbourhood Improvement Programme that received parliamentary approval in 1973 and included 322 local authorities, dealt with reconstruction of old housing, jointly with selective destruction of ramshackle housing. It allocated funds for social and community services, while insuring participation of residents in the decision-making process (Lyon & Newman, 1986; Carter, 1991).

For France, the Neighbourhood Social Development policy announced in 1981, reached 150 neighbourhoods throughout the country. The policy was directed towards comprehensive and integrated management of housing, education, social integration, employment, professional training, health, culture and leisure, with emphasis on involvement of the residents in the processes of change (Tricart, 1991).

The third generation focused on revitalisation, especially in city centres. Revitalisation is a business-like approach emphasising economic development. In the beginning of the 1970s, an economic slow-down was spreading worldwide. Very low prices of land and housing in the city centres began to attract both small and large private entrepreneurs. The new processes can be divided into two groups: public-individual partnerships and public–private partnerships. The first term refers to cases in that investments by individual people, households and owners of small businesses in deteriorated neighbourhoods are supplemented directly through subsidies or indirectly (special regulations, investments in the surrounding public services and so forth) by the authorities. Public–Private Partnership (PPP) refers "…cooperative ventures between the state and private business …" (Linder, 1999: 35).

Methodological Underpinnings of Land Markets

The methodology for assessing land markets is influenced by the need for robust and dynamic data in understanding urban systems since, "planners and decision-makers need to be acquainted with not only the current state of affairs, but also ideas about the future, if they can be able to see the possible consequences of plans and policies they make and recommend" (White & Engelen, 2000: 384). The need to have future knowledge on impacts of policies point to the use of predictive computational models. Since these models capture endogenous instabilities and dynamics in space, they can be referred to as dynamical GIS. Several urban growth modelling approaches have been developed and sporadically reviewed by urban planners, geographers and ecologists (Caputo et al., 2015). One of the common modelling approaches in urban studies since the 1980s engrosses cellular automata (CA). Cellular automata are a system of cells interacting in simple ways but generating complex overall behaviour. "A cellular automaton (A) is defined by a lattice (L), a state space (Q), a neighbourhood template (l) and a local transition function (f) expressed in set notation as A = (L, Q, l, f)" (White & Engelen, 2000: 392).

CA were developed to investigate the behaviour of dynamical systems and many geographical applications still emphasise theoretical issues (White & Engelen, 2000: 398). Such work is necessary, because there remains much to learn about the dynamics of spatial systems. However, CA that demonstrate a capacity for modelling real geographical systems with precision and realism are being increasingly developed. In the future CA-based models may begin to find a place as practical policy and planning tools. On a positive note, the integrated CA modelling approach is able to capture the complex mix of predictability, uncertainty, and novelty in the real world. For this reason, CA models, like any other model, are increasingly adopted as participatory spatial planning tools as they allow for both exploration and visual appreciation of effects of policy proposals that make it easier for planners to communicate the rationale of certain policies to the general public (Silva & Wu, 2012: 139). This is beneficial not only to policy- makers, but also to end users of policy (White & Engelen, 2000). Since the science of CA modelling is still evolving and does not have fixed standards and methodology yet, CA models can also be used to simulate local scale dynamics in the development of housing estates and other land parcels (Batty et al., 1999).

Understanding the distinct factors and physical, ecological and sociological processes that shape land markets in a particular urban system is an important step before modelling (Berling-Wolff & Wu, 2004). This is done through effectively quantifying urban landscape pattern and projecting its spatiotemporal dynamics. This improves applicability of models in real world contexts after they are developed, given the gap in model development in laboratories and their limited application in solving practical urban challenges. Understanding and quantifying drivers of urban system dynamics is followed by scenario building. Scenarios are, "plausible and often simplified descriptions of how the future may develop" (Reginster & Rounsevell, 2006: 635). Scenario building allows explorations of understanding, though it does not provide predictions. The scenarios represent alternative policy visions, and the ways in that each vision could potentially unfold into the future. The scenarios enable local planners to better understand the cost of small-scale (salient) drivers on regional urbanisation, demand for land and the limitations imposed by local planning.

Global Experiences

Urban housing supply from 1995 to 2005 had no effect on the urban land market in China (Chen et al., 2011). The reason was that supply of urban housing in China by then was not decided by the actual demand from the urban housing market, but was controlled by land supply approved by the state. Other than depending on state supply of land, urban housding supply was also shaped by business strategies of real estate development companies. Although China has joined the ranks of housing market nations, its urban housing market, still preserves some statism features including state-controlled land market. The close relationship between local government and land development corporation (LDC) is not a secret in China (Chen et al., 2011: 7–8).

However, the unique result of the urban housing supply suggests that overdependence on the urban housing market as the major provider of housing led to a severe problem in many regions in China. Substantial amounts of land have been left idle by the Land Development Corporation (LDC), rather than being developed to argument housing supply. The hoarding of land in many cities in China is massive. In recent developments, the central government policy that regulates real estate companies provides that developers must develop land within two years, otherwise government repossesses it. This policy exposes the exceptional mechanism of urban housing supply in China that is contrary to the widely accepted economic notions of liberalisation. The Chinese government should give further consideration to its urban housing policy, that regards the housing industry as the major driver of national economic development. An overestimated urban housing market without crucial provision and allowance for public housing may cause serious social problems and damage the national economy. Such lessons were evident in the current subprime financial crisis in a number of Western countries.

In contrast to other countries, migration of Chinese people comprises of two categories based on the status of these migrants. Most rural–urban migrants are excluded from the category of urban residents in official censuses. Moreover, this "special population group is not considered as a part in the urban housing provision system by Chinese housing policy makers" (Chen et al., 2011: 7–8). The emerging urban development patterns in Beijing strongly bear a resemblance to those in other cities with a free market economy (Ding, 2004: 1904). Both land prices and land development intensity fall with distance. Different land uses exhibit different rates of spatial decline in land prices and land development density. This, in turn, drives the spatial separation of land-uses. In other words, office and commercial development have an economic advantage in locating close to the city centre, whereas industrial development is pushed further away towards the suburbs. Residential development is most likely to take place in between. This reflects the strong influence of markets on urban spatial form.

Land rent gradients shifted between 1993 and 2000, reflecting the reduction in transport costs as a consequence of highway construction. The direction of changes in land rent curves led to development of an urban pattern that is common in western cities (Ding, 2004: 1904). Hence, land markets in China are maturing and have an effect on land-use decisions. Maturity of land markets is also reflected by the rise in elasticity of land–capital substitution, suggesting high developers' response to any change in land prices. Like most, if not all empirical studies, the data frequently caused problems.

Developers are always tempted to construct more floor space to increase their profits. This inevitably increases floor area ratios since land input is fixed. In response to this, Beijing Municipal Government imposed heavy fines and punishment to enforce contract execution and control building height since the late 1990s. Another issue is associated with the dual land-allocation system. The dual land-allocation system was developed when the administrative channel of land-allocation remained integral even though public land leasing was spreading throughout the entire nation.

Unlike the Chinese experience, the landscape of urban land development radically changed during two critical periods in the US. From 1949 to 1974, the country witnessed the first round of urban renewal and from 1992 to 2007 another phase occurred (Hyra, 2012). These redevelopment phases, though occurring in different historical contexts, have many similarities including use of federal funds to redevelop minority neighbourhoods near the CBD, displacement of poor minorities, and the movement of concentrated poverty.

These similarities show that cities have been through a new round of urban renewal. Urban policy in the 1950s levelled single-family-style homes that were considered unlivable and built high-rise public housing (Hunt, 2009). During the 1990s, a new round of slum clearance took down these high-rises and erected mixed-income villages that tried to mimic housing stock demolished in the 1950s. While acknowledging the recurring pattern of slum clearance, significant differences between these two redevelopment periods exist. There was the presence of a more intricate set of redevelopment forces and a more complicated set of consequences for African-Americans. Noting the differences between old and new urban renewal is vital for understanding how urban policy in the twentieth and twenty-first century affected metropolitan America.

A hallmark of America's urban land markets is exclusion, often manifested through fiscal and large-lot zoning (Cervero & Duncan, 2004: 299). Municipal zoning favouring commercial, high tax-yielding land uses and high-end housing through minimum lot size requirements has fostered fragmented patterns of urban growth. This, in turn, gave rise to jobs–housing imbalances and spatial mismatches in the location of job opportunities and sites where jobless populations reside. The effects of fiscal zoning and land-use exclusion have been perverse in California. The passage of Proposition 13 in 1978 froze local property tax proceeds to 1% (percent) of assessed land values, unleashing intense competition for high tax-generating land uses.

In the San Francisco Bay Area and Southern California, recent job growth has concentrated in pricey housing markets while most starter-home tracts are being built 50-plus miles away where land is cheap and neighbourhood resistance is minimal. In Atlanta, two visions of urban renewal came into conflict in the mid-1960s (Holliman, 2009). On one side were the Atlanta's mayor, aldermen, city planners, downtown business associations and the African-American middle class. This group advocated for use of federal renewal funding to ensure economic growth for the city, either by protecting central business district property values from nearby slums or by constructing new revenue-generating structures. On the other side were the city's poorer residents, housing advocates, civil rights organisations and neighbourhood activists. Contrary to the mayor's groups, this section advocated for the use of federal renewal funding to replace deteriorating housing stock with new public housing or low-cost housing built by private developers. Based on the competing interest visions for urban renewal, leaders competed for limited funds, site selection, and planning control.

In Punjab Province of Pakistan, well-functioning urban land and housing markets are critical success factors for achieving robust economic growth. However, the Punjab's present markets are not performing adequately as there exists a range of impediments to efficient urban land and housing market performance. These include; excessive public land ownership, inadequate infrastructure services, weak property rights, pervasive public- and private sector rent seeking, counter-productive urban planning policies and regulations, costly sub-division and construction regulations, limited property development financing, rent controls and inadequate property-tax-based revenue-generating mechanisms (Dowall and Ellis, 2009). This shows the diversity of challenges in Punjab's land and property markets.

Focus on Africa

This section discusses issues surrounding the development and dynamics of urban land markets and resilience in African countries. Counties considered include South Africa, Kenya, Zambia and Zimbabwe.

African cities are in shambles because of missing information. The urban plans for Africa's larger cities, if they exist at all, are typically found in a dilapidated state, possibly pinned to the wall in a central government ministry or folded into a large technical report (Watson, 2013). Most often, they reflect a static land use zoning plan covering the older parts of the city and the plans bear little liaison to the reality on the ground. However, this is changing since proposed new urban master plans for many of Africa's larger cities are now to be found on the websites of international architectural, engineering and property development companies. These plans differ from African urban reality than did the post-colonial zoning plans.

Ensuring urban land markets are efficiently managed to achieve economic and social needs is one of the most pressing issues in cities throughout the Third World (Gough & Yankson, 2000). The costs of poor land management are enormous because once land is developed, it cannot easily be reclaimed. The relationship between people and land is often complex and differs between societies depending on history, culture and legal systems. In sub-Saharan Africa, prior to colonialism, policies and objectives of land management were clear. Land was collectively owned whilst individuals only had right of use. The objective was to ensure easy access to land by every household in a predominantly subsistence society. Today, however, customary land tenure operates alongside Western tenurial systems in many urban centres resulting in great ambiguity on land policies and objectives.

Given similar levels of economic development and inequality, it seems strange that no attempt has been made by South Africans to learn from the past experience of urbanisation in countries, such as Brazil, Chile and Colombia (Gilbert & Crankshaw, 1999). South African migrants are becoming more like Latin Americans in demonstrating a growing attachment to the city. Many Sowetans' lives do not fit the model of the 'circular' African migrant; many have lived in Johannesburg for a long time. Semi-permanent residence means that residential movement within the city is increasingly similar to that typical in many Latin American cities (Gilbert and Crankshaw (1999). Judging from Latin American experience, dismantling of formal apartheid systems is unlikely to reduce levels of social segregation in Johannesburg. This is because apartheid remains pronounced in residential property location in terms of who can live where (Bond, 2006). These typical class divisions are just as effective as apartheid in guaranteeing residential segregation.

Such segregation did not end in 1994, but took a class-based character and continued. For example in 2003, Gauteng Province Housing Minister Paul Mashatile then admitted that the resulting landscape had become an embarrassment: 'If we are to integrate communities both economically and racially, then there is need to depart from the present concept of housing delivery that is determined by stands, completed houses and budget spent' (Bond, 2006: 345–346). Similarly, the spokesperson for Gauteng (Zulu) further explained that, 'the view has always been that when low-cost houses are built, they should be built away from existing areas because it impacts on the price of property' (Saturday Star, 7 June 2003). Other than the need to maintain property of high value in urban centres, segregation of the poor in terms of housing location in urban South Africa is also prompted by economic rationales that 'low-cost houses should be developed in outlying areas where the property is cheaper and more houses (can) be built' (Saturday Star, 7 June 2003).This reflects generally-held conceptions on location of low-cost housing not only by the provincial authority but also among real estate companies.

Planning law in RSA is built on a notion that once a township has been established, town-planning schemes determine the use rights that are available to the landowner (Kihato & Berrisford, 2006). These town planning schemes inherited from the pre-democratic era and drawn up in terms of the Provincial Planning Ordinances of the old four provinces of the RSA, govern the historically white, India and coloured urban areas. Much more rudimentary schemes, promulgated under the Blacks (Community Development) Act, 4 of 1984, and the Blacks (Administration) Act, 38 of 1927 govern the historically African urban areas (townships). This has created multiple town planning schemes in urban areas. For instance, the Cape Town municipality has to administer 27 different schemes within its urban area.

In terms of characterising these townships, the term "township" has no formal definition (Pernegger & Godehart, 2007). It refers to underdeveloped, usually (but not only) urban, residential areas reserved for non-whites (Africans, Coloureds and Indians) who lived or worked in areas that were designated 'white only' under the Black Communities Development Act, Sect. 33, Proclamation R293 of 1962, Proclamation R154 of 1983 and the GN R1886 of 1990 in Trust Areas, National Homelands and Independent States). This was mainly during the apartheid era. Today the term "township" is synonymous with neighbourhoods located on the periphery of towns, low-income housing estates and informal settlements. Common characteristics typical of townships include poor community facilities, low commercial investment (shebeens, tuckshops), high unemployment, low household incomes and poverty.

Land Tenure Issues in South Africa

Although formal segregation ended with the new democracy, apartheid policies still have a legacy on the current land and property markets in the townships and low-income housing estates. Denial of land ownership by African residents in urban areas prompted emergence of different tenure systems among formal and informal (renting and subletting) that exist up to today. The rental housing scene for low-income settlements in Cape Town and Johannesburg is characterised by landlords older than tenants, larger families, more ground space for homes and better services similar to those found in other third world cities. Since the renting is informal, rental conditions in townships in Cape Town and Johannesburg are poor and the rental charges are low, so landlords rarely make money. The low rentals accentuate the feeling that is it not worth investing in proper rental accommodation hence most South African landlords merely offer space for tenants to build their own shacks in the backyards.

Other than informal rental markets, private letting and sub-letting are also prevalent. Private letting and sub-letting has existed as a tenure option since the development of townships in the apartheid era, although the practice was illegal. The co-existence of these different tenure systems and sub-markets within most cities creates a complex series of relationships in that policy related to any one tenure system has unintended repercussions on others (Payne, 2000). The influence of land policy on de jure and de facto tenure systems and sub-markets within the townships is characterised by complex interactions. Tenure regularisation programmes operating at city level are likely to reduce market distortions, but impose an excessive burden on land registries (Payne, 2000: 10). Conversely, similar programmes at local level are easier to cope with, but are likely to increase urban land market distortions since a significant proportion of the black urban population in South Africa is in rental accommodation (Gilbert et al., 1997). This negatively challenges policy on development of land markets in South Africa. This negatively affects sustainable development of Cape Town land markets, hence managing spatial development in the townships have been of interest to Neighbourhood Development Partnership Programme (NDPP) (Donaldson et al., 2013).

However, the biggest challenge faced by the NDPP is to encourage municipalities to engage in planning for future-oriented development of these townships. Not only was focus to be given to future infrastructure needs, but municipalities were encouraged to address existing challenges since unaddressed infrastructure requirements today form tomorrow's infrastructure backlogs (Pernegger & Godehart, 2007).

Subsidised, community-controlled social housing (community land trusts, housing associations and cooperatives) may prove to be the only viable approach as financial constraints are some of the huddles to a postmodern urban solution beyond the townships. In Johannesburg, the inner-city neighbourhoods of Joubert Park and Hillbrow are victims of notorious blanket "redlining" (geographical lending discrimination) (Weekly Mail 25 November 1990). Ironically, this, in turn, may foil ambitious redevelopment visions of both the Urban Foundation and Ampros' "Anglo American Inner-City Property Development Fund." An Ampros brochure labels Hillbrow as "undoubtedly one of the primary residential hubs of South Africa that provides investors with excellent potential in terms of both rentals and sales."

In terms of the progress made to reduce social and spatial divisions in South African cities since independence, notable impacts are seen in the distribution of public services in Cape Town. Public services are gradually extending to historically neglected townships (Turok, 2001: 2349). Despite the improvement in service delivery, the character of economic and social development differs markedly across the city. Private-sector investment and jobs continue to be concentrated in the affluent north and west, while low-income subsidised housing is concentrated on the poorer south-east area. Institutional practices and market forces further reinforce these spatial divisions despite the fact that they have costly consequences not only of the poor, but the wider society and the rest of Cape Town's economy.

Concerning challenges of residential property development within the Cape Town province, though the new Unicity Authority provides scope for a more coherent, integrated, proactive and strategic framework to manage land development. This is in the interest of the whole city and of building a stronger employment base in the south-east. In addition, there are conflicts of interests in township redevelopment, between extracting short-term banking profits (economic gains) and prioritisation of human social needs (Bond, 2000). Similarly, other objective constraints on low-income housing provision including exorbitant interest rates, land speculation, and the building materials cartels exist. Each of these constraints exacerbated as South Africa entered the 1990s and under such conditions, modernist or postmodernist commodification of townships is difficult (Financial Mail, 31 May 1991).

The retail sector forms a critical element of a community's economic and social welfare as it provides people with wider choices of goods and services. These choices, until recently, were very limited in township areas (Ligthelm, 2008). For the townships, there are several types of retail establishments in South Africa. These include informal spaza shops prevalent primarily in townships. Typically, these businesses operate in a section of an occupied home or in another structure on a stand zoned or used for residential purposes and where people permanently live. Many of these are run as family entities. The other type of retail properties are tuck-shops. These operate around public transit stations. These informal retailers fill a gap in making goods accessible to people and conveniently service provision in the townships (Rolfe et al., 2010).

Other than tuck-shops another common micro-enterprise found in South Africa's poor areas is the shebeen or informal taverns. These are mostly located in black informal townships—the best-known being in Soweto. Outside Johannesburg, South African shebeens were driven by opportunities for viable markets within the townships (Morris et al. 1997). Other informal retail activities can also be found around taxi ranks and train stations. As elsewhere in Africa, hawkers without a permanent structure are a common feature of retail trade as well. The ubiquitous spaza shops, shebeens, and other micro-enterprises in South Africa face problems including competition from the copycat mentality of entrepreneurs in the same locality. This is a phenomenon common in other developing countries and is, especially, true for street-hawkers (Chan, 2008). Offsetting this negative effect of retail concentration are the positive shopping externalities, or spill-over effects, found in urban areas. Proximity to major population centres could stimulate viability of micro-enterprise in South Africa.

Another issue is that of supermarketisation of urban space. Distribution of supermarkets in Cape Town reflects the model of supermarket diffusion identified in international literature where spatial location of supermarkets in Cape Town is highly stratified according to income (Battersby & Peyton, 2014). As a result, the pre-1994 retail property landscape in low income areas (townships) was characterised by small, often informal businesses offering basic household necessities with poor range of choice. This has resulted in residents preferring to shop outside townships (outshopping). However, rapid income growth of township residents since 1994 turned the townships into lucrative emerging markets for large retail industries. Consequently, national retail supermarket chains have begun to develop formal commercial properties and activities within these townships, resulting in an improvement in retail service and substantial increase in consumer expenditure in these areas, known as 'in-bound shopping'. Whilst development of formal retail activity presents positive changes on social life and economic development of townships, traditional small retail properties and business including spaza/tuck shops are threatened by permanent displacement.

Another case study considered is Mombasa. The case study of land and property developments in Mombasa, Kenya illustrates the general principle that waterfront revitalization schemes cannot realistically be developed in isolation but should be perceived and planned in the context of the wider urban fabric of port cities (Hoyle, 2001). Such cities should be considered as elements within the environmental and management systems of the coastal zones within that they are situated. Besides integration at national level, revitalisation should also be integrating the city in the regional and global urban networks to that they belong. An essential component in Mombasa's Old Town Conservation policy is the reinforcement of historical identity and architectural heritage, refurbishment of buildings, redesign of open spaces at neighbourhood scale and integration of historical and archaeological promenades (Hoyle, 2001). This is derived from key objectives of the policy including modernisation of transport and communication infrastructure, development of mixed land-use, development of human capital to support new economic activities and rehabilitation of the waterfront to boosting urban tourism.

Besides being an essential component of Old Town Conservation Policy, rehabilitation of selected waterfront environments was meant to improve urban tourism and increase public access to the waterfront (Hoyle, 2001). Though the geography of the Old Town poses a natural advantage, successful implementation of these objectives had to be aligned to public and private finance, urban services, administrative skills, cultural tourism and public participation.

In the case of retail development in Zambia, supermarketisation of retail property developments is a common emerging phenomenon (Battersby & Peyton, 2014). However, the major challenge associated with the process in Zambia is "supermarket revolution myopia" (Abrahams, 2010: 116). This is characterised by occurrences of transformation in other sectors of the food market that challenge the dominance of the supermarket sector. Since these transformations are ongoing, the associated implications of commercial property sector remain unclear (Humphrey, 2007).

In Zimbabwe, Bryceson and MacKinnon's (2012) term, 'mineralised urbanisation' as an apt description of Zimbabwe's history of urbanisation and growth of Zimbabwean cities. Mineralised urbanisation refers to the change in urban form and urban profiles arising directly and indirectly from mineral production cycles and related investments and commodity cycles at local, national, regional and global levels (Bryceson & MacKinnon, 2012). During periods of economic growth, mining companies running mining settlements invested in urban infrastructure, education, sports and recreation infrastructure (Tipple, 2000). Following investment primarily for mining purpose, other related activities/ firms, such as retailing, banking, insurance, wholesale and equipment suppliers where attracted to further invest in mining towns. However, since the 1980s, mining firms and activities were negatively affected by government control policy (indigenisation policy) and fluctuating commodity prices following liberalisation in the 1990s. Consequently, mining firms struggled to compete in given high operational costs. As a result, mining investment continued to decline, thousands of workers lost jobs and mining operations eventually closed. This is typical of mining firms specialising in tin, copper, iron and chrome, respectively in Kamativi, Hwange, Redcliff, Zvishavane. Mhangura, Shamva.

However, some of these towns were revived during the commodities boom of the 2000s (Shurugwi, or example), especially those producing platinum metals. Towns, such as Kadoma and Redcliff, that had relied on iron ore production, continued to stagnate. Further structural changes have been erected since the late 1990s have had an impact on growth and land-use development in towns. The decline in corporate-based mining and the closure of some mines have been accompanied by the rise in small-scale and artisanal mining. This was accelerated by the fast track land reform programme that opened up access to more land that was formerly controlled by a few thousand farmers. The land became available for artisanal and small scale mining by both newly resettled farmers and displaced farm workers.

Other than making more land accessible, the other change brought about by the land reform in the urban land markets was the rise of development of informal residential properties in the peri- urban zones of towns. After losing their sources of livelihoods thousands of displaced farm workers informally settled in the peri-urban area. Initially, this occurred in main urban areas, like Harare, Kadoma, and Chegutu but, since early 2000 it has spread to most cities in Zimbabwe.

However, Operation Restore Order in 2005 forced a number of displaced populations residing in urban and peri-urban areas to seek refuge in some of the abandoned mining settlements, occupying the vacant houses and using whatever resources and infrastructure there was. As a result, mining towns whose populations had declined saw an increase in numbers, but not necessarily in economic productivity with a few exceptions. Following decline in mining activities, some mining centres like Chegutu, Kadoma diversified into agro-processing that improved their resilience to economic shocks (cf. Reggiani et al., 2001; Penny & Bryant, 2011; Salat & Bourdic, 2012; Penny et al., 2013).

In addition, other mining towns like Zvishavane have diversified into providing tertiary education facilities – that is typical of university towns in developed countries. Former physical properties in the town are currently being transformed into learning halls and associated properties and activities to support a viable tertiary education environment. These include students' residential quarters, retail facilities and recreational property development (Chronicle, 2015). These cases illustrate positive attempts to prevent deterioration of mining town amid a decline in mining activities though implying an adjustment of the respective local authorities' land management policies.

Despite the positive changes in some of the mining towns, s like Mhangura and Hwange, remained dependant on extraction of a single mineral commodity. Overall, there has been a transition from "prosperous mining towns, to ghost towns then to havens" (Kamete, 2012). Most of the population in these former mining towns survives on petty trading, artisanal mining of gold, tribute mining and casual labour on the resettlement farms. These former mining towns have attracted a growing population of homeless people seeking housing and land where there is infrastructure.

Informal peri-urban developments in cities compromise effective urban transformation. This is because these transitional areas are characterised by insecure tenure and lack of basic services (UN-Habitat, 2010). Adding to these challenges, residential property developments in the peri-urban also suffers from a lack of incorporation into formal urban areas that affect the property values of some of the investments in these areas, repelling meaningful real estate developments. Insecure land tenure arrangements are a constraint to investment; not only for households, but for local authorities as well. This reflects challenges of land institutions to effectively foster sustainable land-use.

Other than poor institutional frameworks for effective land management, the issue of leap frog spatial expansion characterising Harare and other Zimbabwean towns also represent contested interests in management of urban space between ZANU-PF and MDC run urban council where political interest of ZANU-PF party compromise effective management of peri- urban land markets for sustainable development (Muchadenyika, 2016).

Discussion

In most countries in Africa, land markets and land administration institutions are poorly developed as compared to the land management developments in UK and the United States. Failure of local authorities to accommodate changes in emerging economic and human activities in cities by lacking to suitable land at convenient costs and location compromise the economic viability of cities. This is typical of the failure of most former mining towns in Zimbabwe to diversify and allocate land for other activities outside mining amid the deterioration of the mining activities in these towns. Failure to diversify and accommodate new land demands in the market has turned such cities into ghost towns that compromise the resilience of such mining towns. The scenario of deterioration in these towns also signifies lack of the land institutional mechanisms to adjust to new needs for space.

Other than the economy and social dynamics, the development of proper land tenure systems is necessary to improve the resilience of land markets in urban areas (Armitage, 2006; Ahern, 2011; Allan & Bryant, 2011; Anderies, 2014; Barnes & Nel, 2017). This is because it models legal institutionalised relationships of the land-users to the land, the relationship that can be adjusted as per needs in economic, social and environmental context. In addition, a clear tenure system clearly defines how the land users have to relate to the land through legal contracts, a breach of which compromises urban land management and attracts fines. Hence, proper land tenure helps enforce responsibility of households and property developers over effective stewardship of the land (Desoto, 2010).

Conflict of interest in urban land is another challenge to the sustainable management of land in urban areas. In the Zimbabwean case, this emanates from tensions between ZANU- PF ruling party authorities and the MDC dominated urban councils. This brings a conflict of interests and compromises attempts of urban councils to regulate urban land market dynamics. It tends to promote uneven spatial development in urban areas that threatens the sustainability of land management. However, this issue of conflict of interest is not restricted to the case of Zimbabwe alone but is also prevalent in urban revitalisation programmes in the United States of America and in South Africa, particularly under apartheid rule. This shows how equally important political will is for effective transformation of urban land markets.

Conclusion and Policy Options

The study sought to tracc thc history and dynamics in land markets in the global and regional arenas with consideration to resilience and sustainability. This was donc through looking at the changes in demand for land and the dynamics in land policies and management approaches against the dynamics in the economic and social environments. Results reveal that the practices fail to ensure a sustainable urban land markets due to a host of challenges ranging from insecure tenure, failure to adapt to social and economic environment dynamics, lack of political will to institutional challenges where policy that influence land-use changes is often driven by the interests of the elites at the expense of the poor. To develop and maintain effective land markets, several critical factors have to be in place. In addition to clear policies and enforceable laws, the institutional framework needs to be in place for reliable, responsible and accountable decision-making within the general national land policy. The land registration system must also be affordable so that all citizens, especially women and minority groups, rich and poor, can have access to it. Continuous globalisation will inescapably shock land markets, especially as information technology, including web-based services, provides greater access to national land information services.

References

Abhas, J. K., Miner, T. W., & Stanton-Geddes, Z. (Eds.). (2013). *Building urban resilience: Principles, tools, and practice. directions in development*. Washington, DC: World Bank. https://doi.org/10.1596/978-0-8213-8865-5.

Ahern, J. (2011). From fail-safe to safe-to-fail: Sustainability and resilience in the new urban world. *Landscape and Urban Planning, 100*, 341–343. https://doi.org/10.1016/J.Landurbplan.2011.02.021

Allan, P., & Bryant, M. (2011). Resilience as a framework for Urbanism and recovery. *Journal of Landscape Architecture, 6*(2), 34–45. https://doi.org/10.1080/18626033.2011.9723453

Anderies, J. M. (2014). Embedding built environments in social-ecological systems: Resilience-based design principles. *Building Research & Information, 42*(2), 130–142. https://doi.org/10.1080/09613218.2013.857455

Armitage, D. (2006). Resilience management or resilient management? A political ecology of adaptive, multi-level governance. Working Paper Prepared for The IASCP Panel on Community-Based Conservation in A Multi-Level World Held On 19–23 June 2006 In Bali, Indonesia.

Barnes, A., & Nel, V. (2017). Putting spatial resilience into practice. *Urban Forum, 28*(2), 219–232. https://doi.org/10.1007/S12132-017-9303-6

Battersby, J., & Peyton, S. (2014). The geography of supermarkets in Cape Town: Supermarket expansion and food access. *Urban Forum, 25*, 153–164.

Batty, M., Xie, Y., & Sun, Z. (1999). Modelling urban dynamics through GIS-based cellular automata. *Computers, Environment and Urban Systems, 23*, 205–233.

Benenson, I. (1998) Multi-agent simulations of residential dynamics in the city. *Computers, Environment and Urban Systems, 22*(1), 25–42.

Bennett, R., Wallace, J., & Williamson, I. (2008). Organising land information for sustainable land administration. *Land Use Policy, 25*(1), 126–138.

Berling-Wolff, S., & Wu, J. (2004). Modeling urban landscape dynamics: A case study in Phoenix, USA. *Urban Ecosystems, 7*, 215–240.

Bond. (2006). Globalisation/Commodification or Deglobalisation/Decommodification in urban South Africa. *Policy Studies, 26*(3/4), 337–358.

Bryceson, D., & Mackinnon, D. (2012). Eureka and beyond: Mining's impact on African Urbanisation. *Journal of Contemporary African Studies, 30*(4), 513–53714 (Of Malawian and Mozambique Heritage).

Campanella, T. J. (2006). Urban resilience and the recovery of New Orleans. *Journal of the American Planning Association, 72*(2), 141–146.

Caputo, S., Caserio, M., Coles, R., Jankovic, L., & Gaterell, M. R. (2015). Urban resilience: Two diverging interpretations. *Journal of Urbanism: International Research on Placemaking and Urban Sustainability, 8*(3), 222–240. https://doi.org/10.1080/17549175.2014.990913

Carmon, N. (1999). Three generations of urban renewal policies: Analysis and policy implications. *Geoforum, 30*, 145–158.

Cervero, R., & Duncan, M. (2004). Neighbourhood composition and residential land prices: Does exclusion raise or lower values? *Urban Studies, 41*(2), 299–315.

Chen, J., Guo, F., & Wu, Y. (2011). One decade of urban housing reform in China: Urban housing price dynamics and the role of migration and urbanization, 1995–2005. *Habitat International, 35*, 1–8.

Chronicle. (2015). MSU Relocates Faculties. *Chronicle*, 20 July.

Cote, M., & Nightingale, A. J. (2012). Resilience thinking meets social theory: Situating social change in Socio-Ecological Systems (SES) research. *Progress in Human Geography, 36*(4), 475–489.

Coutu, D. L. (2002). *How resilience works*. Harvard Business Review, May 2002.

Davoudi, S., Shaw, K., Haider, J. L., Quinlan, A. E., Peterson, G. D., Wilkinson, C., Fünfgeld, H., Mcevoy, D., & Porter, L. (2012). Resilience: A bridging concept or a dead end? "Reframing" Resilience: Challenges for planning theory and practice. Interacting traps: Resilience assessment of a pasture management system in Northern Afghanistan, Urban resilience: What does it mean in planning practice? Resilience as a useful concept for climate change adaptation? The politics of resilience for planning: A cautionary note. *Planning Theory and Practice, 13*(2), 299–333. https://doi.org/10.1080/14649357.2012.677124.

De Weijer, F. (2013). *Resilience: A Trojan Horse for a new way of thinking*? (ECDPM Discussion Paper 139). Maastricht: European Centre for Development Policy Management. (http://www.ecdpm.org/dp139).

Du Plessis, C. (2008). Understanding cities as social-ecological systems. In *World Sustainable Building Conference – SB'08*, Melbourne, Australia. 21–25 September. http://researchspace.csir.co.za/dspace/handle/10204/3306.

Ding, C. (2004). Urban spatial development in the land policy reform era: Evidence from Beijing. *Urban Studies, 41*(10), 1889–1907.

Donaldson, R., Du Plessis, D., Spocter, M., & Massey, R. (2013). The South African area-based urban renewal programme: Experiences from Cape Town. *Journal of Housing and the Built Environment, 28*(4), 629–638.

Dowall, D. E., Ellis, D. (2009). Urban land and housing markets in the Punjab, Pakistan. *Urban Studies, 46*(11), 2277–2300.

Du Plessis, C. (2012). Towards a regenerative paradigm for the built environment. *Building Research & Information, 40*(1), 7–22.

Du Plessis, C., & Cole, R. J. (2011). Motivating change: Shifting the paradigm. *Building Research & Information, 39*(5), 436–449. https://doi.org/10.1080/09613218.2011.582697

Du Plessis, C., & Brandon, P. (2015). An Ecological worldview as basis for a regenerative sustainability paradigm for the built environment. *Journal of Cleaner Production, 109*, 53–61.

Elmqvist, T. (2014). Urban resilience thinking. *Solutions, 5*(5), 26—30. http://www.thesolutionsjournal.org/node/237196?page=16 22/2/2016.

Ernstson, H., Van Der Leeu, S. E., Redman, C., Mefferet, D. J., Davis, G., Alfsen, C., & Elmquvist, T. (2010). Urban transitions: On urban resilience and human-dominated ecosystems. *Ambio, 39*, 531–545. https://doi.org/10.1007/S13280-010-0081-9

Filatova, T., Parker, D., & Van Der Veen, A. (2009). Agent-based urban land markets: Agent's pricing behavior, land prices and urban land use change. *Journal of Artificial Societies and Social Simulation, 12*(1, 3), 1–24.

Gilbert, A., & Crankshaw, O. (1999). Comparing South African and Latin American experience: Migration and housing mobility in Soweto. *Urban Studies, 36*(13), 2375–2400.

Gilbert, A., Mabin, A., McCarthy, M., & Watson, V. (1997). Low-income rental housing: Are South African cities different? *Environment and Urbanization, 9*(1), 133–148.

Gough, K. V., & Yankson, P. W. K. (2000). Land markets in African Cities: The case of Peri-Urban Accra, Ghana. *Urban Studies, 37*(13), 2485–2500.

Holliman, I. V. (2009). From crackertown to model city? Urban renewal and community building in Atlanta, 1963–1966. *Journal of Urban History, 35*(3), 369–386.

Hoyle, B. (2001). Urban renewal in East African Port Cities: Mombasa's old town waterfront. *GeoJournal, 53*, 183–197.

Hyra, D. S. (2012). Conceptualizing the new urban renewal: Comparing the past to the present. *Urban Affairs Review, 48*(4), 498–527.

Kamete, A. Y. (2012). Of prosperity, Ghost Towns and Havens: Mining and Urbanisation in Zimbabwe. *Journal of Contemporary African Studies, 30*(4), 589–609.

Kihato, M. & Berrisford, S. (2006). *Regulatory systems and making urban land markets work for the poor in South Africa. Urban landmark Position Paper 4.* Paper Prepared for The Urban Land Seminar, November 2006, Muldersdrift, South Africa. Lemanski, C.

Klein, B., Koenig, R., & Schmitt, G. (2017). Managing Urban resilience. Stream Processing Platform for Responsive Cities. *Informatik-Spektrum,40*(1), 35–45.

Landman, K., & Nel, D. (2012). Reconsidering urban resilience through an exploration of the historical system dynamics of two neighbourhoods in Pretoria. In *Paper Presented at the Planning Africa Conference*, 17–19 September 2012, Durban, South Africa.

Lemanski, C. (2014). Hybrid gentrification in South Africa: Theorising across Southern and Northern Cities. *Urban Studies, 51*(14), 2943–2960.

Ligthelm, A. A. (2008). The impact of shopping mall development on small township retailers. *SAJEMS, 11*(1), 37–53.

Linder, S. H. (1999). Coming to terms with the public-private partnership: A Grammar of multiple meanings. *American Behavioral Scientist, 43*(1), 35–51.

Maclean, K., Cuthill, M., & Ross, H. (2014). Six attributes of social resilience. *Journal of Environmental Planning and Management, 57*(1), 144–156.

Martin-Breen, P., & Anderies, J. M. (2011). *Resilience: A literature review*. Bellagio Initiative Partners: Institute of Development Studies (IDS), The Resource Alliance and The Rockefeller Foundation. http://opendocs.ids.ac.uk/opendocs/handle/123456789/3692#.vs3wk_j96m8 22/2/2016.

Mcphearson, T., Andersson, E., Elmqvist, T., & Frantzeskaki, N. (2015). Resilience of and through urban ecosystem services. *Ecosystem Services, 12*, 152–156.

Meerow, S., Newell, J. P., & Stults, M. (2016). Defining urban resilience: A review. *Landscape and Urban Planning, 147*, 38–49.

Muchadenyika, D. (2016). Land for housing: A political resource-reflections from Zimbabwe's urban areas. *Journal of Southern African Studies, 41*(6), 1219–1238.

Nel, D. H., & Nel, V. J. (2012). *An exploration into urban resilience from a complex adaptive systems perspective*. Paper Presented at The SAPI Planning Africa Conference On "Growth, Democracy and Inclusion: Navigating Contested Futures", Held in Durban On 17–19 September 2012.

Nijkamp, P., Van Der Burch, M., & Vindigni, G. (2002). A Comparative Institutional Evaluation of Public–Private Partnerships in Dutch Urban Land-Use and Revitalisation Projects. *Urban Studies, 39*(10), 1865–1880.

O'Hare, P., & White, I. (2013). Deconstructing resilience: Lessons from planning practice. *Planning Practice & Research, 28*(3), 275–279. https://doi.org/10.1080/02697459.2013.787721

Payne, G. (2000). *Urban land tenure policy options: Titles or rights?* Paper Presented at The World Bank Urban Forum, Westfields Marriott, Virginia, USA, 03–05 April, 2000.

Pernegger, L., & Godehart, S. (2007). *Townships in the South African Geographic Landscape—Physical and social legacies and challenges*. Pretoria: Training for Township Renewal Initiative' (TTRI).

Reggiani, A., De Graaff, T., & Nijkamp, P. (2001). *Resilience: An evolutionary approach to spatial economic systems*, Tinbergen Institute Discussion Paper, No. 01–100/3.

Reginster, I., & Rounsevell, M. (2006). Scenarios of future urban land use in Europe. *Environment and Planning B: Planning and Design, 33*, 619–636.

Rolfe, R., Woodward, D., Ligthelm, A., & Guimarães. (2010). *The viability of informal micro-enterprise in South Africa*. Paper Presented at The Conference On? Entrepreneurship in Africa? Whitman School of Management, Syracuse University, Syracuse, New York, April 1–3.

Salat, S., & Bourdic, L. (2012). Systemic resilience of complex urban systems: Of trees and leaves. *Tema, 2*, 55–68. https://doi.org/10.6092/1970-9870/918

Seeliger, L., & Turok, I. (2013). Towards sustainable cities: Extending resilience with insights from vulnerability and transition theory. *Sustainability, 5*, 2108–2128. https://doi.org/10.3390/Su5052108

Seeliger, L., & Turok, I. (2014). Averting a downward spiral: Building resilience in informal urban settlements through adaptive governance. *Environment and Urbanisation, 26*(1), 184–199.

Sellberg, M., Wilkinson, C., & Peterson, G. D. (2015). Resilience assessment: A useful approach to navigate urban sustainability. *Ecology and Society, 20*(1), 43–66.

Shaw, K. (2012). "Reframing" Resilience: Challenges for planning theory and practice. *Planning Theory & Practice, 13*(2), 308–312. https://doi.org/10.1080/14649357.2012.677124. Retrieved on 4 April 2015.

Silva, E., & Wu, N. (2012). Surveying models in urban land studies. *Journal of Planning Literature, 27*(2), 139–152.

Stumpp, E. (2013). New in town? On resilience and "Resilient Cities." *Cities, 32*, 164–166.

Tipple, G. A. (2000). *Extending themselves: User-initiated transformations of government built housing in developing countries.* Liverpool: Liverpool University Press.
Turok, I. (2001). Persistent polarisation post-apartheid? Progress towards urban integration in Cape Town. *Urban Studies, 38*(13), 2349–2377.
UN-Habitat. (2010). *Urban land markets: Economic concepts and tools for engaging In Africa.* Nairobi: UN-Habitat.
Van Niekerk, W. (2013). Translating disaster resilience into spatial planning practice in South Africa: Challenges and champions. *Jamba: Journal of Disaster Risk Studies, 5*(1), 53, 6 Pages. https://doi.org/10.4102/jamba.v5i1.53
Walker, B., & Salt, D. (2006). *Resilience thinking: Sustaining people and ecosystems in a changing world.* Island Press.
Walker, B. H., Gunderson, L. H., Kinzig, A. P., Folke, C., Carpenter, S. R., & Schultz, L. (2006). A handful of heuristics ADN some propositions for understanding resilience in social-ecological systems. *Ecology and Society, 11*(1), 13. [Online] http://www.ecologyandsociety.org/vol11/iss1/art13/
Watson, V. (2013). African urban fantasies: Dreams or nightmares? *Environment & Urbanization, 26*(1), 215–231.
White I., & O'Hare. (2014).From rhetoric to reality: That resilience, why resilience, and whose resilience in spatial planning? *Environment and Planning C: Government and Policy, 32*(5), 934–950. https://doi.org/10.1068/C12117.
White, R., & Engelen, G. (2000). High-resolution integrated modelling of the spatial dynamics of urban and regional systems. *Computers, Environment and Urban Systems, 24*, 383–400.
Wilkinson, C. (2012). Social-ecological resilience: Insights and issues for planning theory. *Planning Theory., 11*, 148–169. https://doi.org/10.1177/1473095211426274
Wilkinson, C. (2012). Urban resilience: What does it mean in planning practice? *Planning Theory & Practice, 13*(2), 319–324. https://doi.org/10.1080/14649357.2012.677124.
Woolf, S., Twigg, J., Parikh, P., & Karaoglou, A. (2016). Towards measurable resilience: A novel framework tool for the assessment of resilience levels in slums. *International Journal of Disaster Risk Reduction.* https://doi.org/10.1016/j.ijdrr.2016.08.003.
Williamson, I. P. (2001). Land administration "'Best Practice"' providing the infrastructure for land policy implementation. *Land Use Policy, 18*, 297–307.
World Bank. (2012b). *Climate change, disaster risk, and the urban poor: Cities building resilience for a changing world.* Washington, DC: The International Bank for Reconstruction and Development/The World Bank. (https://openknowledge.worldbank.org)
World Bank. (2013). *Why resilience matters: The poverty impacts of disasters* (Policy Research Working Paper 6699). Washington, DC: The International Bank for Reconstruction and Development/The World Bank. (https://openknowledge.worldbank.org)
Wu, J., & Wu, T. (2013). Ecological resilience as a foundation for urban design and sustainability. *Resilience in ecology and urban design* (pp. 211–229). Springer.

Innocent Chirisa is a Full Professor in Environmental and Regional Planning at the University of Zimbabwe. At the present time, he is the Dean of the Faculty of Social and Behavioural Sciences at the University of Zimbabwe. He is the Deputy Chairman of the Zimbabwe Ezekiel Guti University. He has served as the Deputy Dean of the Faculty and Chairman of the Department of Rural and Urban Planning at the same University. His keen interest is in urban and peri-urban dynamics. Currently focusing on environmental systems dynamics with respect to land-use, ecology, water and energy. Professor Chirisa is a seasoned scholar who also contributes in a local Sunday Mail newspaper.

Chapter 2
The Many Faces to Resilience and the Practical Implications to Urban Development and Management in Zimbabwe

Tinashe N. Mujongonde-Kanonhuhwa and Innocent Chirisa

Abstract This chapter examines the many faces to resilience and the practical implications to urban development and management in Zimbabwe interrogating aspects of places, people and processes. It emanates from the observation that urban vulnerability is on the increase in the country and beyond, taking various forms including physical, socio-economic and environmental. Methodologically, the chapter is built on critical literature review and case study of the different urban areas (mega, meso and micro). The results indicate that Zimbabwe's urban areas are characterised by many stresses and shocks that branch from the natural, technological or socio-economic factors, some of which, may threaten animal and human livelihoods and the existence of cities. It appreciates that the African landscape is sometimes characterised by unique challenges that require intensive research and appreciation of local knowledge through public engagement in specifically dealing with them. It also analyses the role that government has to play in promotion of resilient-city building through policing and interlinkages with other nations.

Introduction

It is projected that by 2050, over half of the world's population will be living in cities. In the face of increased urbanisation and changing urban dynamics, the urban landscape now faces large risks from flooding, rising sea levels and other natural hazards, such as unbearable extreme weather conditions and very high temperatures. Poor urban planning practices have been pointed out as major contributors of the increased vulnerability of urban areas. This calls for city planners, designers, architects, and many experts to form a roundtable and share ideas in producing resilient city landscapes that are capable of absorbing shocks ad stresses to maintain the smooth functioning of urban systems.

T. N. Mujongonde-Kanonhuhwa · I. Chirisa (✉)
Department of Demography Settlement & Development, University of Zimbabwe, Mt Pleasant, PO Box MP167, Harare, Zimbabwe

I. Chirisa and A. Chigudu (eds.), *Resilience and Sustainability in Urban Africa*, Advances in 21st Century Human Settlements,
https://doi.org/10.1007/978-981-16-3288-4_2

Urban resilience is defined as "the ability of the city region, (i.e. its institutions and population) to prepare for, react to, and recover from sudden shocks and long-term disruptions and to maintain its central functions when exposed to these hazardous events" (Schiappacasse & Müller, 2015: S13). The urban sphere is now very vulnerable and today's planning of cities must take note of the various threats to urban systems, hence the need to ensure that adaptive and mitigation strategies are in place. It is often argued that, the term resilience is more popular in areas of disaster risk reduction and climate change adaptation (Graham et al., 2016). This may partly be as a result of the increased prevalence of disasters (both natural and man–made) and the climatic changes that are currently being experienced globally.

Theoretical Underpinnings

This chapter unpacks theories that relate to urban vulnerability and resilience, community building and resilience and urban management. It also aims to establish the various links between these, and how they can be comprehensively looked at when planning and designing for resilient cities.

The concept of resilience is broad and holds a number of theoretical assumptions that state that it is heavily influenced by race, gender, and economic status, whilst being affected by the availability of environmental resources (Greene 2002 in Greene et al., 2004). It stresses the ability of people and urban environments to cope with life stress (Greene et al., 2004). Resilient urban environments should therefore be able to withstand stresses posed by the natural, political and socio-economic environments. This requires urban planners and managers to have a proper understanding of cultural diversity and religious affiliations of a population before taking action against a specified problem. Urban resilience strategies are influenced by power differentials and also affected by availability of resources. Resource availability determines decisions to be made in terms of prioritising areas to focus on when pushing for resilient building and development of cities.

A community has been referred to as the brain of the city, as it directs activities, responds to its needs and learns from its experience (Godschalk & Xu, 2015). An ideal community resilience is the ability of a community to respond and recover from disasters. A community is an organ that must absorb stresses and shocks with minimal or no disturbances in case of any disasters. It is then necessary to uniquely define a place/community/city as this would help to model resilient urban practices in a way that suits the local people. This defines the African nations as unique, that requires the need to study each landscape to find local solutions that work. This, in turn, makes it easier for urban management practices to be put in place, in a way that would not promote resistance to change.

The world is now a global village, thus there is need to understand how global changes have affected urban context and its institutional landscape (Corubolo, 1998). Understanding the urban changing dynamics helps to come up with urban management practices that work well for the urban system as a whole. Urban management covers a wide area and these include enabling the private sector through regulatory framework to manage urban settlements, reform of public policy to encourage efficiency, transparency and accountability as regards to state bureaucracy and management of infrastructure and urban delivery through public sector management (McCarney et al. 1995 in Corubolo, 1998). Urban management practices are therefore aimed at efficiently directing urban processes and functions to allow the smooth running of urban cities. City managers are left with a task to manage city functions in such a way that limits the impact of urban vulnerability impacts.

Van Dijk and Kwarteng (2007), stress the importance of Local Government to have performance targets and accountability benchmarks. The need to reach performance targets and later become accountable for the results becomes a critical element in promoting good governance. In executing good urban management practices, it is critical for the planning body to understand the risks threatening urban areas, and examines the nature of the various communities that make up a city (their needs and culture, for example), to come up with well-planned resilient cities that are able to withstand the shocks of various stresses and disasters.

Literature Review

Urban vulnerability takes many forms and it is the responsibility of City Planners and Managers, Governments, Communities and engagement of people at grass-root levels to put ideas together and effectively prepare and design cities in such a way that they can effectively mitigate / positively absorb effects of hazardous events to maintain functionality.

The urban sphere has increasingly become very vulnerable throughout the years and is now experiencing shocks from both the natural and man-made phenomena. The urban sphere can suffer from sudden shocks and long-term disruption as caused by natural, technological, political and socio-economic factors. Natural factors may include droughts, wildfire, earthquakes, floods and storms, whilst technological factors may come in the form of explosions, chemical spills and radiation (United Nations 2015; in Schiappacasse & Müller, 2015). Socio-economic factors may include housing crisis, social and political conflicts, and wars (United Nations, 2015 in Schiappacasse & Müller, 2015). It is quite evident that the urban sphere is pregnant with many uncertainties that stem from many corners, some of which threaten the existence of cities. Urban vulnerability is filled with so much complexity, that calls for the government and policy-makers to come up with decisions and well-funded policies that aim at producing resilient cities capable of positively dealing with urban shocks. To guard against the destruction of cities as a result of unpreparedness and failure to prevent or cope with these shocks, urban spaces must become more resilient.

Resilience is the ability of a system to absorb disturbances, recover, and reorganise while retaining the same structure, functions, and feedbacks after suffering a major perturbation (Saavedra et al., 2012). In order for the urban system to effectively function well, governments must invest and support city planning and designing and redesigning of new and old cities to come up with urban areas that can withstand any form of technological or natural shocks. Urban resilience strategies may include urban agricultural practices, green infrastructure, early warning systems for tropical storms, earthquakes and floods and resilient housing.

Urban Agriculture as a Way to Promote Urban Resilience

Urban agriculture is a dynamic concept that hints on livelihood systems ranging from subsistence production and processing at household level, to fully commercialised agriculture (Sebata et al., 2014). Urban agriculture, although practiced in both developed and developing economies, often serves different purposes, such as recreation in the former and food security in the latter (Pearson et al., 2011). This is mainly because, African economies rely on subsistence farming as a way to feed the family. Urban agriculture can be practiced on micro, meso and macro scales. Individual household may practice urban agriculture by planting in their backyards and creating green roofs. At meso-scale, community gardens and urban parks are created and at macro level, commercial scale farms and greenhouses may be utilised (Pearson et al., 2011).

Furthermore, urban agriculture aims at promoting community-based adaptive management by creating a diversified pool of agricultural produce, being a source of technological innovation and learning for the urban poor (Zeeuw et al., 2011). This will not only enhance the technical skills of the local people, but reduce poverty as it becomes possible for people to feed themselves. Urban dwellers are presumed to buy most of their food, and food security depends on whether a household can afford buying, that also relies on the household's disposable income (Kutiwa et al., 2010). In times of economic crises however, people's disposable incomes would shrink, and urban agriculture may become an important activity to feed the urban population, especially the poor.

Climate Change and Urban Resilience

Climate change is now at the forefront of international policy agenda due to its ability to cause interstate/intrastate conflicts (Chagutah, 2013). This indicates the need for climate change to be managed well to promote good relations between nations and locally. Mitigation and adaptive strategies must be put in place to allow for future city development in a way that promotes climate change adaptability. Understanding the urban system and how it is affected by urban climatic changes, and fully considering the vulnerable groups that fall prey to these changes becomes essential in effective city development and management.

Analysing the various components of the city and how they work, and the direct and indirect impacts of climate change on urban systems helps in planning and designing for the built and soft environment to minimise the effects. Understanding the different vulnerable groups helps to identify the most vulnerable and those who can respond to shocks and stresses in case of any. This kind of information assists in making plans that may be easier to implement in a quest to promote formation of resilient cities.

In solving issues, such as droughts and flooding, municipalities and urban councils ought to be creative and learn from other nations on how they are managing. Floods can be viewed as an opportunity or a threat, depending on level of preparedness. An example is that of Friedrikstad in Norway, that has planned for the maximum use of storm water by practicing an integrated approach to storm water management through implementation of the municipal master plan of 2008–2018 (Nie, 2018). This shows the level of seriousness on the part of the municipality through adoption of innovative ways aimed at capturing the benefits from storms and floods that sometimes occur in their environment.

It is argued that urban African spaces are highly diverse, and most of them are shaped by colonial legacies and increased urbanisation practices (Zievorgel et al., 2017). Most of the African landscapes are characterised by a lot of urban inequality, poverty and informality. Harare is no exception as it is the hub of street vending, transport and housing informality, and invasion of urban space through urban agricultural practices (Dube & Chirisa, 2012). In resilient city building, one must understand the different and various land uses that characterise the area before planning and designing solutions for particular problems. This helps in creating solutions that last, thus, preventing the use of blue-print plans in solving African city problems.

In preparing for disaster emergencies, the government of Zimbabwe passed the Civil Defence Act soon after independence in 1982, and later the Civil Protection Act of 1989, where the state would channel all available national resources through the establishment of the National Civil Protection Fund in case of emergency (Chikoto and Sadiq, 2012 in Bongo et al., 2013). This was meant to be a responsive mechanism to guide the government and organisations to respond in times of serious crises. The disaster response system is open to international and regional support and includes the input from traditional leaders (Chatiza, 2019). In case of failure by government to solve the crisis however, it would then declare a state of emergency and seek for funds even from the international community.

The Act was later amended in 2001 (Bongo et al., 2013). The legislation culminated in the formation of the Department of Civil Protection that operates under the Ministry of Local Government, Rural and Urban Development (Bongo et al., 2013). It can be argued that the disaster and risk management plan of Zimbabwe considers local knowledge since there is also the involvement of traditional leaders who are assumed to represent the interests of the locals. The system is comprehensive as it allows the sourcing of funds, ideas and human resources from the government, public and private organisations and the international community.

In efforts to solve urban challenges, African urban centres appreciate the use of Geographical Information Systems (GIS) in solving some of the urban challenges and Zimbabwe is no exception. GIS can be used to solve some of the most challenging urban problems, such as traffic congestion, urban sprawl, lack of infrastructure, environmental deterioration and map the risk of natural disaster occurrence in certain areas (Mabaso et al., 2015). This would help in urban management practices by controlling how cities grow and take shape. A city may also be prevented from deteriorating as it can be used to identify areas needing an upgrade. Using the software to identify potential areas needing attentions, can help planners and the government to allocate resources accordingly based on priority areas.

Research Methodology

The research methodology was based on the archival method of data collection, where data was to be analysed from ready-made sources/cases. In a similar study on climate resilient cities, cities to be studied were selected on the basis of city size, data availability and classification as either climate active or climate inactive (Saavedra et al., 2012). In this study, selected cities of Harare, Bulawayo and Mutare were used as case examples on the basis of size and information availability. The three cities represent the three largest cities in Zimbabwe. The zones have in one way or the other been affected by climate changes and natural disasters, such as Cyclone Idai that hit the Manicaland region heavily. This section is based on a case-study approach as a way to learn how these areas have tackled various shocks and how cities have managed to grow and become more resilient throughout the period. Data was collected through the archival method by comparing various literature from different authors.

Results

This section discusses the various efforts if any that have been made by Zimbabweans to prepare, cope, and mitigate potential hazards through resilience planning and building of cities. Results are discussed by referring to the three cities, of Harare, Bulawayo and Mutare that are under study.

Harare

It is quite evident that Zimbabwe's urban areas are vulnerable to many expected and unexpected shocks. As previously discussed, the Zimbabwean urban spaces, particularly Harare, is characterised by high levels of informality, urbanisation and lack of clean water to drink, among other challenges. Some of the shocks observed in Harare's high-density of Tafara, for instance, included the cholera outbreak of 2008, the typhoid outbreak of 2009, the Diarrhoea outbreak of 2014 and the closure of local industries from 2006 to 2010 (Pharoah, 2016). The occurrence of such events threatened human lives and livelihood that further disturbed the effective functioning of the economy.

These mishaps did not affect Tafara high-density alone, but also affected other locations in Harare, such as Budiriro, Glenview and Mabvuku residential areas, thereby threatening the effective functioning of the health system and the economic performance of the country as many cases were reported. In a survey carried out, most people believed that these problems were as a result of lack of water, sanitation and hygienic practices that had been due to the lack of services, such as clean piped-water, unemployment and food security (Pharoah, 2016). In a bid to solve the inadequate water saga, the government with the help of organisations, such as UNICEF facilitated the drilling of boreholes in the high-density areas of Budiriro, Glenview, Tafara and Kambuzuma. In another study, Tawodzera (2010), stresses that most households in Harare are food insecure with the majority failing to afford a single meal per day. This has resulted in urban agricultural practices by the urbanites as a way to obtain cheap food.

The City of Bulawayo and Urban Agriculture

Sebata et al. (2014), acknowledges the benefits of urban agriculture enjoyed by women in Bulawayo's high-density area of Cowdray Park. These include the creation of employment, generation of income and nutritional improvement. Besides providing food security, urban agriculture also helps in improving the people's livelihood (Kutiwa et al., 2010).

The practice of urban agriculture also helped to regulate urban temperatures and curb the effects of floods and droughts, such as hunger. This, in turn, upgraded the people's standards of living as food security was guaranteed. However, the disconnection between urban agriculture and city ecology was due to concentration by Planners on the built environment, and less focus by researchers on urban agriculture (Pearson et al., 2011). There is need for Planners to comprehensively plan for both the soft and hard landscapes to ensure connectivity of both systems. Government support in terms of financial resources and regulatory framework to legalises urban agriculture becomes important.

Mutare

The Manicaland Province, especially Chimanimani, Chipinge and Mutare, was hit by the catastrophic flooding disaster of Cyclone Idai in March, 2019, leaving 340 people dead and thousands missing, at the same time, destroying 250 boreholes, 13 health centres, 489 tobacco barns and 40 dip-tanks, among others (Chatiza, 2019). This not only shows destruction of people's livelihoods, but also damage to infrastructure. However, Cyclone Idai managed to reveal the extent of disaster preparedness of the country and the extent of their responsiveness and that of NGOs and other local and international organisations. Apart from government, the situation was assisted by donor-funded organisation, churches, the United Nations and ordinary citizens, that shows that people and organisations were able to come together to achieve the common goal of saving lives and helping the affected people living in the area.

Lessons Drawn and Discussion

It is important to come up with local solutions and adaptive mechanisms in creating resilient cities and solving some of the challenges in Zimbabwe today. Zivoergel et al. (2017) acknowledges that African urban spaces are characterised by a mixture of formal and informal links, diverse knowledge and practices hence, the need to provide opportunities for building resilience using the bottom up approaches. Involving the local people through a participatory approach also helps build local capacity and preparedness and anchors sustainable (hazard-resilient) development (Chatiza, 2019). This shows the importance of involving the locals as it creates a strong base for current and future preparedness of hazardous events, in case of any.

Often, disaster risk reduction has also been modelled along the needs and priorities of able-bodied people, whilst largely excluding those with various forms of impairments (Bongo et al., 2013). Chatiza (2019) stresses the need for government to invest in adaptive and resilient measures to protect women and other vulnerable groups, and encourage the setting up of social and child protection systems that are sensitive to disaster. This would not only save lives, but also empower local people to be aware and reach out in case of any possible disasters. Involving people at all levels would also help reduce resistance to change and by so doing, people may work together towards the common good.

Zivoergel et al. (2017) explain the importance of placing urban resilience within global systems and that African cities need to be recognised as part of a neoliberal era of global finance, capital accumulation and global circuits of communication. This helps in fostering support within different regions as nations help each other since they would view processes and places and the environment as a system that completes and hence not compete with each other.

Moreover, from events of the occurrence of Cyclone Idai; unanticipated natural disasters, state capacity to predict and prepare for the occurrence of such natural hazards has become critical (Chatiza, 2019). This shows the importance of the government not only to lobby for funds to buy equipment to be able to forecast such events, but also the need for a nation to be equipped with the necessary skills that would help in ensuring a proper disaster risk assessment of the area. There is need for local capacity building even in universities to equip students with skills necessary to make positive contributions in case of any disaster threats. Also, it is important for a nation to maintain good relations with other countries regionally, even at international scales to get assistance in the event of a disaster.

In creating resilient cities of greatness; governments, policy-makers and designers ought to work together to come up with solutions aimed at mitigating and adapting to vulnerable urban conditions being experienced today. Strategies to cope with climate changes, for instance, include water-harvesting and green infrastructure and resilient housing. Green infrastructure can be used as a vital element in blending the built and the natural environments together, as it is attached to a lot of benefits. Green Infrastructure promotes health and recreational wellbeing improves the land and property values of the area, and maximises the area's aesthetics.

Green infrastructure practices will not only lessen the burden of erratic climate changes on the environment, but also help to solve problems related to traffic congestion, such as pollution, food production and security and replenishment of groundwater sources. However, these may be expensive to implement, but cheaper to maintain in the long run. It is, however, important for Urban Designers and Planners to view issues in a holistic manner such that they fully understand the problems likely to be created as they try to solve another problem. In trying to solve the problems of traffic congestion through road expansion, Green Infrastructure is reduced and habitat connectivity is damaged (Schiappacasse & Muller, 2015). Thus, in solving one issue, another one with future negative effects may arise, that calls for the need to fully evaluate the different outcomes of choosing a solution over the other, perhaps the increased use of public transport would have been better than road expansion in certain scenarios.

Inadequate planning practices also have the ability to cause future disasters. The issuing out of stands and building of houses in Harare suburbs, such as Tafara, has resulted in the uptake of wetlands and dried riverbeds, and infills that increase the risks of floods in the area (Pharoah, 2016). Other examples may include Belvedere West that is constructed on wetland soils. Also, the spread of unprotected wells also threatens the outbreak of waterborne diseases (Pharoah, 2016). This has also been a common phenomenon to other new residential suburbs in Harare South, such as Southlands, Stoneridge and Hopley high-density residential areas. To avoid hazardous events, as a result of poor planning, there is need for planning authorities to be strict in terms of controlling development as a way of positively shaping city development and, in turn, avoid future hazards from forming as a result of lack of proper planning practices.

Resilience building has increasingly gained importance with it being mainly used to deal with issues as regards to urban risk management. The global North has shaped many policies relating to the resilience agenda. However, the applicability of these policies in African countries has not been sufficiently assessed (Ziervogel et al. 2017). There is need to clearly evaluate each system and come up with uniquely designed policies that work as per specified problematic area. Ziervogel et al. (2017) also advocate for the implementation of negotiated resilience to include diverse interests when making policies. The global South must therefore consult various stakeholders from the grassroot levels to come up with home-grown solutions to solve some of the problems.

Urban resilience must be well placed in global systems to allow challenges to be viewed holistically and allow African nations to contribute based on local and regional experiences (Ziervogel et al., 2017). This is mainly because issues such as climate change and urban heat islands may have their effects in Africa, whilst the real causes are deeply rooted in the global north. It is argued that the gospel about green infrastructure has no planning status, and has limited success as regards to practical implementation, since its success depends on the extent of it being embedded in comprehensive urban and regional planning approaches (Schiappacasse & Müller, 2015). This calls for serious consideration and participation on the part of the government and Planners to come up with written down policies to be effected when planning our cities.

Zimbabwe's city planning system does not cater for urban agriculture (Kutiwa et al., 2010). It is still regarded as an informal activity (Sebata et al., 2014). In promotion of urban agricultural practices, it is important for policy-makers to come up with a policy and regulatory framework that lays down parameters and criteria within that urban agriculture can be carried out to prevent urban farmers from playing hide and seek with city authorities (Tawodzera, 2010). This would eliminate fear in people as they get security that the urban farming practices are perfectly legal and can freely contribute to a family's food security. Kutiwa et al. (2010) however, acknowledge the Nyanga and Harare declarations on urban agriculture made by Ministers of Local Government in Eastern and Southern Africa, that helped that urban agriculture contributes to urban food security, local economic development and poverty reduction (Hungwe, 2006 in Kutiwa et al., 2010). The declaration paved way to formulation of policies and legal frameworks for urban agriculture, such as the National Environmental Draft policy that local authorities can use as a tool to plan on how to integrate urban agriculture in planning (Kutiwa et al., 2010). This shows positive improvements in promotion of urban agriculture, however, there is need for stern policy measures. There is also need for capacity building through tertiary education to train people on how to use technology to be able to forecast the possible natural disaster, that would allow a country to prepare for and reduce the effects of the hazardous event.

Conclusions and Policy Alternatives

The chapter acknowledged that the world now suffers from many shocks and stresses, some of which depend on their location in the global North or South. It has been noted that, African cities, particularly Zimbabwe, do suffer from unique challenges emanating from the global North. They include increased urban informality in the form of informal housing, traffic congestion and street-vending. Other challenges include droughts, floods, and unbearably high temperatures. It was discovered that each city has its own challenges as regards to shocks that threaten the effective functioning of urban systems and people's wellbeing. It however, discovered that governments and councils are sometimes unclear as regards to promotion of activities that support the creation of resilient cities, such as urban agriculture, that represents a gap between what is known by the public and that which is expected by urban councils. Correspondingly, there is a gap between theories and practice as regards to implementation of green infrastructural practices, that sometimes lack funding as they are expensive to implement, hence the need for government and international funding. This chapter recommends:

- formulation of unique strategies tailor-mode for each environment to come up with solutions that work in producing resilient cities. For instance, there is need to be clear on policy and regulatory frameworks in support of urban agriculture in Zimbabwean cities. This would foster food security and an improved standard of living for the urbanites
- Public participation is also vital as it contributes to creating solutions that are implementable and accepted by the public.
- There is need for government support in terms of funding and capacity development even from primary up to tertiary levels to train the population on various ways on how to handle different types of disasters, to be ready and avoid being caught unaware when disaster strikes.
- The teaching of Geographic Information Systems becomes important to equip people with the skills to be able to map potential threats and challenges that may affect a particular area.
- It is important to note that unique solutions suited for each environment are important. However, it shall not be forgetting that problems also need to be viewed with 'global lenses'. This explains a situation where the world is viewed as a system to understand how activities in one area either affect or get affected by activities in the other. Making such considerations would help solve problems comprehensively.

References

Bongo, P. P., Chipangura, P., Sithole, M., & Moyo, F. (2013). *A rights-based analysis of disaster risk reduction framework in Zimbabwe and its implications for policy and practice. Jàmbá: Journal of Disaster Risk Studies, 5*(2), 1–11.

Chagutah, T. (2013). *Preventing and resolving future climate and natural resource-related conflicts in the zambesi basin: A study of Bulawayo and Chinde Districts.* South African Institute of International Affairs (SAIIA) Occasional Paper Number 155. Available online: africaportal.org.

Chatiza, K. (2019). *Cyclone Idai in Zimbabwe: An analysis of policy implications for post-disaster institutional development to strengthen disaster risk management.* Available online: http://repo.floodalliance.net/jspui/handle/44111/3259

Corubolo, E. (1998). *Urban management and social justice.* Available online: https://discovery.ucl.ac.uk/id/eprint/33/1/wp92.pdf

Dube, D., & Chirisa, I. (2012). The informal city: Assessing its scope, variants and direction in Harare. *Global Advanced Research Journal of Geography and Regional Planning, 1*(1), 016–025.

Godschalk, D., & Xu, C. (2015). Urban hazard mitigation: Creating resilient cities. *Urban Planning International, 30*(2), 22–29.

Graham, L., Debucquoy, W., & Anguelovski, I. (2016). The influence of urban development dynamics on community resilience practice in New York City after superstorm sandy: Experiences from the lower east side and the rockaways. *Global Environmental Change, 40*, 112–124.
Greene, R. R., Galambos, C., & Lee, Y. (2004). Resilience theory: Theoretical and professional conceptualizations. *Journal of Human Behavior in the Social Environment, 8*(4), 75–91.
Kutiwa, S., Boon, E., & Devuyst, D. (2010). Urban agriculture in low income households of Harare: An adaptive response to economic crisis. *Journal of Human Ecology, 32*(2), 85–96.
Mabaso, A., Shekede, M. D., Christa, I., Zanamwe, L., Gwitira, I., & Bandauko, E. (2015). Urban physical development and master planning in Zimbabwe: An assessment of conformance in the city of Mutare. *Journal for Studies in Humanities and Social Sciences, 4*(1 & 2), 72–88.
Nie, L. (2018). Enhancing urban flood resilience—A case study for policy implementation. *ICE themes flood resilience* (pp. 79–92). London: ICE publishing.
Pharaoh, R. (2016). *Strengthening urban resilience in African cities: Understanding and addressing urban risk*. ActionAid International.
Saavedra, C., Budd, W. W., & Lovrich, N. P. (2012). Assessing resilience to climate change in US cities. *Urban Studies Research, 2012*, 1–11.
Schiappacasse, P., & Müller, B. (2015). Planning green infrastructure as a source of urban and regional resilience-towards institutional challenges. *Urbani Izziv, 26*, 13–24.
Sebata, N., Mabhena, C., & Sithole, M. (2014). Does urban agriculture help improve women's resilience to poverty? Evidence from low-income generating women in Bulawayo. *IOSR Journal of Humanities and Social Science, 19*(4), 128–136.
Tawodzera, G. (2010). *Vulnerability and resilience in crisis: Urban household food insecurity in Harare, Zimbabwe.* Doctoral Dissertation. Cape Town: University of Cape Town.
Van Dijk, M. P., & Oduro-Kwarteng, S. (2007). Urban management and solid waste issues in Africa. In *Contribution to ISWA World Congress September, Amsterdam.*
Zeeuw, H., Veenhuizen, R., & Dubbeling, M. (2011). Foresight project on global food and farming futures. The role of urban agriculture in building resilient cities in developing countries. *Journal of Agricultural Science, 149*, 9–16.

Tinashe Natasha Kanonhuhwa is a former Teaching Assistant in the former Department of Rural and Urban Planning, University of Zimbabwe. She holds a Master of Science and a B.Sc. (Honours) in Rural and Urban Planning, University of Zimbabwe. Her research interests are in housing, city development and poverty alleviation and food security.

Innocent Chirisa is a Full Professor in Environmental and Regional Planning at the University of Zimbabwe. At the present time, he is the Dean of the Faculty of Social and Behavioral Sciences at the University of Zimbabwe. He is the Deputy Chairman of the Zimbabwe Ezekiel Guti University. He has served as the Deputy Dean of the Faculty and Chairman of the Department of Rural and Urban Planning at the same University. His keen interest is in urban and peri-urban dynamics. Currently focusing on environmental systems dynamics with respect to land-use, ecology, water and energy. Professor Chirisa is a seasoned scholar who also contributes in a local Sunday Mail newspaper.

Chapter 3
Disaster Management Capabilities in Zimbabwe: The Context of Africa Agenda 2063

Godfrey Chikowore, John David Nhavira, Terence Motida Mashonganyika, and Constantine Munhande

Abstract Historically, disasters have adversely affected nations across the world, inflicting wide ranging losses on one hand while on the other hand creating development opportunities for urban communities. Recovery instruments and transformative efforts by Zimbabwe and international stakeholders in disaster management assumed the form of conventions on climate change and other means commonly agreed at international level that were destined to address perceived humanitarian crisis (Hyogo and Sendai Disaster Frameworks, 2015–2030). Yet, even as disaster presupposes a disruptive situation it equally provokes an attitude of restoration as Zimbabwe reclaims its position in a heavily contested world. Recovery should take the form of industrialisation or infrastructure rehabilitation where economies have been disrupted as Zimbabwe; among nations, makes frantic efforts to align with global development trends. In the light of preceding conversation, this chapter seeks to explore disaster management capabilities of Zimbabwe evaluating hazards and opportunities characteristic of disaster recovery phases in a context of opportunities availed by development cooperation programmes as Africa Agenda 2063. Informed by the theory of disaster management and transformation, the work is founded on a descriptive research design augmented by quantitative and qualitative data analysis and comparative data analysis. In conclusion, the study recommends cooperation in disaster management as a strategy for minimisation of losses on one hand and uplifting the affected urban settlements to ranks of modern global cities as afforded by robust industrialisation programmes grounded in development cooperation frameworks.

G. Chikowore (✉) · T. M. Mashonganyika · C. Munhande
Faculty of Arts, Midlands State University, Zvishavane Campus, Zimbabwe

J. D. Nhavira
Faculty of Business Management Sciences and Economics, University of Zimbabwe, Harare, Zimbabwe

I. Chirisa and A. Chigudu (eds.), *Resilience and Sustainability in Urban Africa*, Advances in 21st Century Human Settlements,
https://doi.org/10.1007/978-981-16-3288-4_3

Introduction

Disaster management capabilities and institutions in Zimbabwe and across regions of the world presuppose that catastrophes historically have adversely affected nations across the world inflicting wide ranging losses on one hand while on the other hand creating development opportunities for urban communities. What this scenario presupposes is that the loss, recovery and transformation strategy adopted by the nation should be very receptive and closely monitor situational development or behaviour with as much accuracy to guarantee marginal loss of human life, property, infrastructure and productive time as is possible. Characteristically anthropogenic and natural in terms of their origin, disasters provoke commensurate management approach that could inherently in principle have four key phases namely; preparedness, response, recovery and mitigation. Generally, preparedness has been pronounced key to successful disaster management if any nation is to avert disastrous scenario where heavy losses in human life, property, infrastructure and productive time have been inflicted. National disaster management capabilities become meaningful not only by themselves but functionally well in conjunction and complementary utilitarian connection with other sovereign states on a series of hierarchical levels. Consequently, recovery instruments and transformative efforts by Zimbabwe and international stakeholders in disaster management should assume not only the form of institutions and conventions on climate change but also other means as commonly agreed on international level specifically to address perceived anthropogenic or natural disaster generated humanitarian crisis. Equally, constructive and progressive recovery; elimination-minimisation or depth of all-round losses and the magnitude of socio-economic cultural transformation in the aftermath of a disaster constitute a reflection of how the windows and opportunities availed by the national, international conventions and institutions have been exploited or ignored by the government of Zimbabwe.

In their requisite Charters the United Nations (UN); Southern Africa Development Cooperation (SADC); the African Union (AU) and the Nation (specifically Zimbabwe) have common positions on strengthening national capabilities for disaster management on assumption of the humanitarian nature of crisis generated by a striking disaster. By interpretation, this means that the degree of receptiveness and self-imposed alienation of a state or Zimbabwe relating to the common charters also reflects on the state disaster management capability level. Yet, even as disaster presupposes a disruptive situation, it equally provokes an attitude of restoration and renewal as Zimbabwe reclaims its position in a heavily contested world. Dialectically understood, the restoration or renewal as a process inherently suggests a definite level of disaster management capabilities for Zimbabwe that should essentially combine the local and external components as determined by the peculiarity of Zimbabwe as a sovereignty on one hand and also as an integral component of the fast-moving global village on the other hand. Recovery or restoration should take the form of industrialisation or infrastructure rehabilitation where economies have been disrupted as the country makes frantic efforts to align with global development trends. As alluded to in the preceding discussion, this chapter seeks to assess disaster management capabilities of Zimbabwe highlighting hazards and opportunities characteristic of disaster recovery phases in a context of opportunities availed by development cooperation programmes as Africa Agenda 2063.

Subscribing to the disaster related provisions of the United Nations (UN); African Union (AU) and the Southern Africa Development Community (SADC), Africa Agenda 2063 is founded on seven key principles of which the most critical to this discussion is: Aspiration 1, that propounds a prosperous Africa based on inclusive growth and sustainable development. Under Aspiration 1 Africa is expected to participate in global efforts for climate change mitigation that support and broaden the policy space for sustainable development on the continent on one hand while on the other hand, addressing the global challenge of climate change. This is achieved by prioritising adaptation in all our actions, drawing upon skills of diverse disciplines with adequate support (affordable technology development and transfer, capacity building, financial and technical resources) to ensure implementation of actions for survival of most vulnerable populations, including islands states, and for sustainable development and shared prosperity (Agenda Africa 2063, 2015).

Greatly augmenting the African Union provisions on upliftment of institutional capabilities are the United Nations Sustainable Goals, especially Goal 1. Its key position states the need by 2030 to build the resilience of the poor and those in vulnerable situations and reduce their exposure and vulnerability to climate-related extreme events and other economic, social and environmental shocks and disasters. It seeks to create sound policy frameworks at the national, regional and international levels, based on pro-poor and gender-sensitive development strategies, to support accelerated investment in poverty eradication actions. Within the context of the above referred windows and opportunities availed by the United Nations and Africa Agenda 2063, Zimbabwe as either a receptive or a non-receptive state will stand to either develop and strengthen or weaken its disaster management capabilities. Not only do the windows allow for bolstering of disaster management capabilities but also guarantee resilient infrastructure that prompts sustainable economic growth in the developing world. Informed by the theory of disaster management and transformation, the work is founded on a descriptive research design augmented by quantitative and qualitative data analysis and comparative data analysis research methods. The study benefits from key contributions by Christopherson (2009), Magaya (2017), Kegley and Shanon (2011), Edgar Pieterse (2008), Alexander (2016), Claire Gillespie (2018), Zimmermann and Stössel (2011), Van der Waldt (2013), Jha, Miner and Geddes (2012).

Literature Review

Several insightful works have highlighted the great set-backs and breakthroughs experienced by nation states across the sub Saharan Africa to that Zimbabwe is an integral component (Alexander, 2016).

A key contribution entitled "Quick facts: Cyclone Idai effects on Southern Africa" exposing the underdevelopment of disaster management capabilities of Zimbabwe and many other nations in Southern Africa, noted that "almost 3 million people in southern Africa had been adversely affected by Cyclone Idai, (Mercy Corps Report, 2019). In Mozambique, close to 200 people had died and more than 1400 were injured. More than 17,000 houses were totally destroyed, displacing thousands of families. In Zimbabwe, more than 344 deaths and hundreds of missing people had been reported, mainly in Chimanimani, especially Machongwe Area, Ngangu, Kopa and other districts, such as Buhera, Makoni, and Mutare, the Skyline and Chipinge Hospitals. About 270,000 people had been affected by flooding and landslides and around 90,000 families were in need of shelter assistance. In Malawi, more than 840,000 people were impacted, with 56 deaths and 577 injuries. More than 94,000 people were estimated to be displaced". Technically disaster management capabilities for Zimbabwe need to show organisation of the institution from the local level typically on ward level, to district, through to provincial and national level where it should link up with the international institutions as provided by the conventions since this certainly determines the nature and level of response and recovery in the event of a disaster strikings.

While making clarification of fundamental principles, national objectives, citizenship, declaration of rights, the executive, the legislature, elections, judiciary and the courts. The 2013 Zimbabwean Constitution conspicuously relegates disaster management issues. Even as disaster greatly relates to fundamental human rights and the all-round security disposition of a citizen, there is hardly any reference to disaster (be it anthropogenic or natural in essence) through the requisite constitutional sections with the same gravity as independent commissions supporting democracy and other sections. This glaring setback makes the probability of existence of effective disaster management institutions a very remote issue that may only be talked of in the event of emergencies. Such a scenario grossly reflects on the limited capacity of Zimbabwe in socio-economic, cultural and scientific terms in the event of a disaster.

Contemporary disaster management debate has been raised in an incisive entitled "City Futures-Confronting the crisis of urban development" (Pieterse, 2008). Contributions accentuate concern about the current "shelter for all" and "urban good governance" policies that have been observed to treat only the symptoms and not causes of the problem. Under such circumstances there stands the urgency of reinvigorating civil society in cities to encourage radical democracy, economic resilience, social resistance and environmental sustainability that is folded into the everyday concerns of marginalised people (Pieterse, 2008). Shelter for all and urban good governance, resilience and environmental sustainability are by implication and interpretation central to disaster management capabilities any nation possesses to the extent that all ultimately relate to human security, socio-economic cultural disposition in a fast-developing global village. Inherently the edition either directly or indirectly reminds about the need for governments to build and develop institutional capabilities that guarantee security, negligible loss in the event of unforeseen developments that may be disastrous or controllable.

Making a logical argument that accentuates the need for urgency in strengthening disaster management capabilities of nations, especially in a fast-changing global population, cities and urban economies and environment is the contemporary work (Stutz & Warf, 2012). On one hand, by logical interpretation and implication for the urgency of effective disaster management, the edition discusses global population distribution and its sustainability through avoiding overpopulation, resources depletion and promoting global cities and urban economies. Resilient urban infrastructure that is disaster resistant to guarantee safety for urban population, industry and agriculture remains a necessity Frederick Stutz and Barney Warf (2012). This warrants the need to strengthen disaster management capabilities and their requisite institutions in the event of anthropogenic and environmental emergencies. Their argument informs on the essence of well-constructed disaster management capabilities and resilient institutions in developed nations where security of swelling urban population, growing cities and urban economies environmental sustainability are well managed within the main overall security guaranteed. On a development and progressive principle, the instances for existence of pronounced disaster management capabilities as implied by resilient infrastructure for cities and urban settlement in developed nations set a precedent for Zimbabwe, the sub Saharan Africa and the rest of the developing world as to how best resilient socio-economic cultural infrastructure should be developed for a guarantee of long-term security and safety of human life, assets, infrastructure and industry.

Considerably complementing the disaster management capabilities in Zimbabwe debate is the UN High Level Panel Report entitled "A New Global Partnership: Eradicate Poverty and Transform Economies through Sustainable Development. United Nations Sustainable Development Goals" (UN 2015). Clearly in definition of the set of goals for the sustainable development and transformation of the world nations with everyone on board, of which Zimbabwe is part, the most relevant are Goals 9, 11 and 13 that thematically accentuate the essence of resilient infrastructure. In Goal 9 the Report emphasises the urgency of building resilient infrastructure, promotion of inclusive and sustainable industrialisation and fostering innovation; while Goal 11 pronounces the making of cities and human settlements that are inclusive, safe, resilient and sustainable. Goal 13 of the Report concludes the debate on effective disaster management institutions by emphasising an urgent adoption of action to combat climate change and its impact. Goal 13 confirms the significance of the signatory status of Zimbabwe and any other nation bound to the requisite international conventions and institutions on combating effects of climate change, natural and anthropogenic emergencies for the guarantee of safety, security and sustainable development. It calls for proper consumption and production patterns as inequality is reduced within and among countries. Within given possibilities on the claim of the commonwealth and value from international conventions these goals justify the case for Zimbabwe in strengthening its disaster management institutions and consolidating its capabilities as a guarantee of safety and security of human life; industrial and domestic infrastructure.

“Disaster Management Cycle” an enlightening contribution by Corina Warfield (2009) is pertinent to the debate on disaster management capabilities of contemporary Zimbabwe, not only makes a concise qualification of the goals of disaster management, but, proceeds to define the phases of disaster management as a well calculated process on averting high deaths levels, and destruction caused by natural disaster. While equally matching the significance of each phase, the further account emphasises the urgency accorded to the preparedness phase before response phase in the event a disaster will have struck.

The Hyogo Framework of Action; a 10-year plan adopted by the World Conference on Disaster Reduction held in Kobe, Hyogo Japan, on January 2005 by 168 governments, was meant to make the world a safer place from natural hazards. This framework was subsequently succeeded by the adopted Sendai Framework for Disaster Risk Reduction 2015–2030 by UN Member States on 18 March 2015. Zimbabwe’s efforts on strengthening disaster management capabilities greatly get a boost from the main purpose of these instruments that also function as a management tool to help countries develop disaster risk reduction strategies, make risk-informed policy decisions and allocate resources to prevent new disaster risks.

The Zimbabwe Civil Protection Unit Act of 1989 was to be replaced in 2015 by the Emergency Preparedness and Disaster Management Act that was meant to effect a more elaborate mechanism for disaster risk reduction. Meant to benefit and derive value for Zimbabwe from the Hyogo and Sendai Disaster Management Frameworks, the 1989 Civil Protection Unit Act cum Emergency Preparedness and Disaster Management Act indicated that disaster preparedness capabilities of Zimbabwe were quite grossly off the mark and expectation as it was literally overwhelmed by the 2019 March Cyclone Idai disaster that among many literally swept away homes, destroyed urban centres, schools, hospitals, cattle kraals and extensively destroyed infrastructure disconnecting places within Chimanimani and Inyanga; and between these zones and the rest of Zimbabwe.

Disaster preparedness capability of Zimbabwe indicates that the country has not fully exploited the valuable potential on enhancement of its capabilities as availed by the Hyogo and Sendai Disaster Frameworks. Even with the presence of the erratic Zimbabwe Civil Protection Act and Unit cum Zimbabwe Emergency Preparedness and Disaster Management Act 2015, the strengths remain grossly overwhelmed by the weaknesses. These weaknesses are exacerbated mainly by a visibly declining socio-economic cultural standing at home and abroad.

Methodology

The study is founded on the theory of disaster management, good governance; development, aid and transformation. In terms of its investigative strategy, the contribution is founded on a qualitative analysis research design. Logically, the whole argument is built on the qualitative design by extensively employing the descriptive and comparative data analysis, qualitative data analysis, illustrative modelling methods of research in combination. Very critical is the philosophical foundations of the study that is in the constructivist school of thought where the core concern is on social groups whereby through shared ideas on disaster management, images and identities develop as change that influences the world outlook (Kegley & Blanton, 2011: 47). Within this philosophical approach, key actors are individuals, intergovernmental organisations and transnational networks due to their abounding social groups with shared ideas, meaning, images and persuasion that brings new understanding. Their underlying principles are very central to the construction of this study. In terms of approach to peace under constructivism, stakeholders promote progressive ideas, while encouraging states to adhere to norms for appropriate and commonly conceived behaviour. Prospects in global terms on disaster management hinge on the prevailing ideas and values, especially the commonly perceived and conceived Hyogo Disaster Management Framework cum Sendai Framework for Disaster Risk Reduction 2015–2030.

The expected outcome of this study constitutes the proposed development of a disaster management capabilities model for transformation in Zimbabwe, founded on the international conventions while taking cognisance of the Africa Agenda 2063 provisions. Certainly, the envisaged capabilities model is meant to ensure recovery, minimise losses and induce a consistent urban transformation in step with global trends on disaster management and post-disaster restoration and reclamation operations with a great bias on industrialisation. Based on the experiences with international conventions and the current national capabilities on disaster management, new approaches on reclamation of disaster struck cities and zones will experience recovery from the point of view of industrialisation and a speedy growth based on combined efforts from the point of view of constructivism.

Discussion

A very lucid clarification of the status of disaster management capabilities in Zimbabwe pointing on the recovery capacity, magnitude of losses and degree of urban transformation was reflected by the two fundamental tragic experiences namely the 2019 Cyclone Idai disaster and the 2014 Tokwe Mukosi Floods Victims (Mudzingwa, 2015; Tau et al., 2016; Zhou et al., 2011). In the absence of consistent implementation of the terms as commonly adopted in Hyogo Disaster Management Framework that subsequently was elevated to be the Sendai Framework for Disaster Risk Reduction 2015–2030, Zimbabwe has shown great setbacks in its systems and institutions on disaster management. Disaster management challenges for Zimbabwe reflect that even at local community, district, provincial through to national level there is still a longer way to go for development of the disaster management systems despite opportunities availed through the Hyogo and Sendai Disaster Management Framework. Essentially by their nature they greatly predispose member states, in particular Zimbabwe, to a great leap-frogging opportunity on transforming the disaster management capabilities from local community, district, and provincial through to national level. In an enlightening contribution by Mavhura (2016) on significance of the Hyogo Framework for Action 2005–2015 Priority 1, an analysis of the Zimbabwe contemporary disaster legislation namely Chapter 10:06 of the 1989 Civil Protection Act, establishes the strengths and weaknesses of community disaster resilience that led to great revelations of the hurdles characterising the disaster management systems and the situation as a whole on national level. Effectively the study established that the 1989 Civil Protection Act fell short of building national and community resilience to disasters. It further established other key weaknesses, such as inactive community participation in Disaster Risk Reduction (DRR), unavailability of dedicated and adequate resources to implement DRR programmes, centralisation of power and resources, and the focus on 'natural' hazards, rather than on vulnerability and resilience.

In concluding the discussion, Mavhura (2016) recommended the urgency for revision of the Act so that it resonates with international best practices on disaster legislation. Even as these revelations were made, there appeared to be a passive effort on development of disaster management institutions on the part of Zimbabwe government as attempts were made by the same government to adopt another Act in a campaign launched by the Department of Civil Protection of Zimbabwe with assistance of the United Nations Development Program (UNDP) in 2005 (Xinhua Report, 2005). This effort entailed replacement of the 1989 Civil Protection Act with an Emergency Preparedness and Disaster Management Act (EPDM) that purportedly differed from the former Act on the strengthened funding requirements; decentralisation of disaster management, integration of early warning systems for emergencies and disasters. This then envisaged EPDM was meant to facilitate the change of name for the Department to an Emergency Preparedness and Disaster Management organisation (EPDM) whose major function included developing risk reduction strategies to minimise vulnerability to both natural and man-made hazards (ibid.).

Notwithstanding these efforts launched in 2005 to create more sophisticated disaster management capabilities in Zimbabwe, there was a degeneration of the situation confirming that the effort made earlier remained theoretical or a biblical mere goat idol as confirmed fourteen (14) years down the line in the Editorial Comment of the Online *Sunday Mail* of March 17.2019. This edition carried the title coded "Time government revisits Civil Protection Act" that act was only fourteen years down the line being examined but only to be referred to or footnoted in a main national newspaper just because a crippling natural disaster had struck in whose preparedness phase the Government of Zimbabwe was hardly traceable. Further commenting on the Cyclone Idai disaster and response of the Government of Zimbabwe, the Sunday Mail Editorial Comment noted that,

> At the time of going to Press, 31 lives had been lost while 80 were still missing in Chimanimani and Chipinge due to the Tropical Cyclone Idai that hit Manicaland Province hard over the past two days. Chimanimani was literally cut off from the rest of the provinces as most bridges were swept away. The havoc-wrecking windy and foggy conditions compounded the situation for rescue teams that found it difficult to access affected areas. While the weather inclement is understandable given the nature of the storm, questions have once again been raised about the country's state of preparedness in times of disasters.

Although Zimbabwe has since 2015 registered significant progress in meeting the requirements of the Sendai Framework for Disaster Risk Reduction that seeks to boost the countries' resilience in the face of disasters, a lot still needs to be done to prevent the loss of lives. Judging from the manner in that the Civil Protection Unit reacted to the Tropical Cyclone Idai, there is literally a lot that still needs to be done. Preparedness and responsiveness are separate parts of disaster management and as a country, we seem to focus more on the latter, rather than the earlier. The country surely needs to be proactive and prepare for disasters most of which would have been predicted. We surely cannot wait for a disaster to happen and then put in place reactionary measures. This scenario denotes absence of elementary preparedness even on village to local level, let al.one district, provincial and national level as the government waited for the disaster to happen and then religiously proceeded to put in place marginal, politicised reactionary measures. Circumstances surrounding the responsive phase of both the 2019 Cyclone Idai and even earlier on the 2014 Tokwe Mukosi Floods Displaced Victims Resettlement visibly point to a grossly fragmented, incoherent and scarcely funded national disaster management capabilities, institutions and systems.

The adversity and magnitude of this institutional fragmentation and lack of funding is manifest, among other key factors, not only in the absence of a sophisticated modern disaster management technical base but also absence of skilfully-trained personnel; absence of pre-disaster community relocation strategies; preparedness strategies, death and a life as usual culture on the part of the community and state authorities; absence of complementary and coherent elementary early warning alert systems on local, district, provincial through to national level; absence of well-equipped early warning holding field evacuation camps catering for needs of all age groups in a human cantered fashion. Above all, the long-term adversity in fragmentation of Zimbabwe disaster management capabilities is shown by dedicated funds from the donor community as confirmed by an Online World News in that the United Nations under the humanitarian aid programme not only requested but further raised the required amount for a recovery from the effects of the Cyclone Idai disaster. In an article entitled "UN raises aid appeal for Zimbabwe to 331.5 US$ mln, many face starvation" prompted by the Cyclone Idai disaster by virtue of its quantum manifested how the Zimbabwe government not only lacked resources. It reveals problems emanating from a heavily incapacitated disaster management institution usually created over the years as shown by the futile process of repealing the 1989 Civil Protection Act to the Emergency Preparedness and Disaster Management Act (EPDM); that to date, has in both theory and practice remained quite obscure.

It is crucial to note that challenges with the long overdue transition from the 1989 Civil Protection Act (CPA) to an envisaged Emergency Preparedness and Disaster Management Act (EPDM) have inherently created fundamental hurdles in the consolidation of a vibrant national disaster management institution that asserts itself timeously in preparedness, responsiveness; recovery and mitigation phases in the event of an impending disaster or after a disaster has struck.

Hyogo Framework and Sendai Frameworks Disaster Management

Commonly adopted by member states as they mobilise efforts and resources on guaranteeing human safety and asserting protection in the event of an anthropogenic or natural disaster eruption, these international and intergovernmental conventions potentially change the theoretical and practical complexion of national disaster management capabilities. With this change, a systematic strengthening of disaster management capabilities should follow on all levels from the ward, district, and province to national level that conveniently project into the regional, continental and global ranks through commonly adopted conventions. Within this kind of appreciation, the constructivists philosophical principles informing this study will most probably find wider expression, as this multilevel approach provides several social groups notably individuals, nongovernmental organisations and transnational networks the essential flexibility to develop shared images and identities on effective disaster management through promotion of progressive ideas and encouragement of states to adhere to proper disaster management norms for appropriate behaviour that would most probably guarantee minimum loses in the event a disaster (Kegley & Blanton, 2011).

Within the Commonly adopted Hyogo Framework of Action (HFA), value added to Zimbabwe disaster management institution involve substantial reduction in disaster losses in lives, and in the social, economic, and environmental assets of communities and countries. The same framework demands the development of progress reports reflecting local, national and regional progress in the implementation of disaster risk reduction actions and establish baselines on levels of scores achieved in implementing the HFA's five priorities for action. The five HFA priorities are: Monitoring and Progress Review process on facilitating the self-assessment of disaster risk reduction measures at the national; regional and global levels; promotion of United Nations Disaster and Risk Reduction efforts; organisations and civil society contributions towards the humanitarian crisis; and finally reporting on the progress of HFA implementation, establishing and strengthening National Platforms, and others, (Prevention Web, 2019). By measure of the ineffectiveness of Zimbabwe disaster risk reduction capabilities, this instrument was not well exploited by the government of Zimbabwe, notwithstanding the potential for greater transformation of the national disaster management system on all levels as alluded to in the preceding argument.

Nevertheless, another greater opportunity for the development of the Zimbabwe Disaster Risk Reduction Capabilities (ZDRRC) stands within a commonly agreed international convention on disaster risk reduction and management concluded on March 18 2015, and adopted by United Nations dubbed the Sendai Framework for Disaster Risk Reduction 2015–2030 (SFDRR). Functionally tallying well with the Goals 1; 11 and 13 of the United Nations with value derived by member states of which Zimbabwe is, the SFDRR destined to make the world safer from natural disasters saddles on seven (7) targets.

Inherently, promoting disaster risk reduction efforts in a responsive Zimbabwe are the following Sendai Framework for Disaster Risk Reduction (SFDRR) 2015–2030 seven global principles namely: mortality rates reduction in global population by 2030 from 2015 high rates; reduction of disaster affected people in 2015 in global population by 2030; reduction of economic loses in global GDP ratios from 2015 by 2030; reducing damage to critical infrastructure and disruption of basic services by 2030; increasing countries with national and local Disaster Risk Reduction management (DRR) strategies; increasing international cooperation with developing nations in DRR operations; increase availability and access to multi-hazard early warning systems and disaster risk information and assessments (Prevention Web, 2019). Viewed within the philosophical paradigm of constructivism, the SFDRR principles unlike the realist and liberal philosophical approaches where war and security, institutionalised peace are the core concern with states being key players, the former perspective imparts greater value through promotion of progressive ideas and encouragement of states to adhere to commonly agreed disaster risk reduction norms as the key players, namely social groups share meanings, ideas and images on effective disaster management (Kegley & Blanton, 2011).

Concluding this discussion, it would be critical to note the opportunity open to Zimbabwe for adoption of the Emergency Preparedness and Disaster Management Act (EPDM) considered a strategic departure from the 1989 Civil Protection Act. Transformative opportunities are quite abundant as these principles well connect with the UNSDGs and the Africa Agenda 2063 principles on promotion of industrialisation and deepening regional integration and cooperation in the face of impact from climate change. Effectively, there is greater scope not only for multi-level development of disaster risk reduction capabilities but more so there exist real opportunities for a simultaneous realisation of an industrialisation process as Zimbabwe contributes more meaningfully to both the national and global economy.

Resilience Through Disaster Management Phases and Deepening Partnerships with Developed World Regions

The Africa Agenda 2063 and United Nations Sustainable Development Goals context of consolidating and transforming the disaster management capabilities in Zimbabwe should be well grounded in a very clear perception and understanding of disaster management cycle or phases namely: preparedness, responsiveness, recovery and mitigation (Corina Warfield, 2009). With each of the phases equally weighing in significance and more important than the other, the value appropriately and timeously derived from each phase does matter in the manner and degree to that it enables saving of human life, infrastructure and assets and wildlife. Depicted in Fig. 3.1, are the two constituent sides of disaster management that Zimbabwe has to be critically aware of namely the "risk management" and "crisis management" aspects of disaster management capabilities development process.

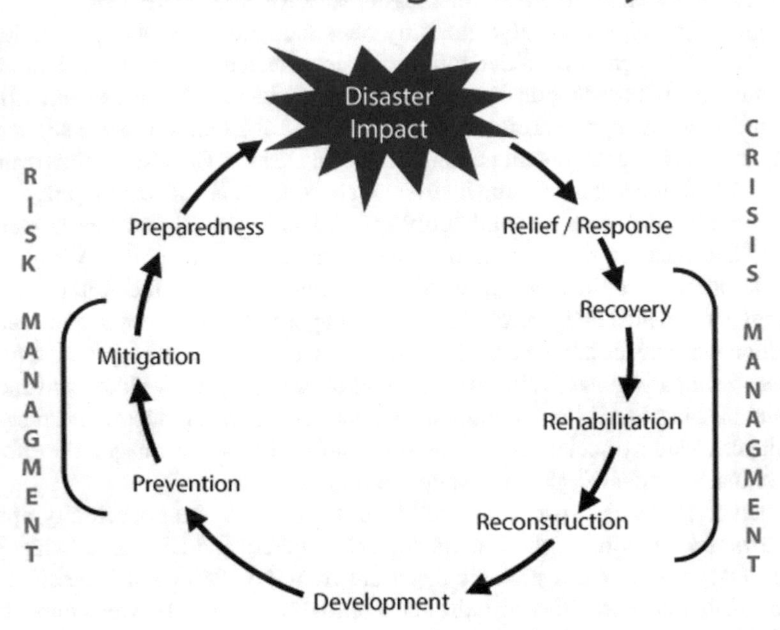

Fig. 3.1 Disaster management phases and strengthening risk reduction capabilities. *Source* Municipal Corporation of Greater Mumbai. Disaster Management Department

The pre-disaster impact phase comprises the risk management side that is made up of three elements namely prevention, mitigation and preparedness, Fig. 3.1.

Understood in theoretical and practical terms Africa Agenda 2063 Aspiration One (1) and United Nations Sustainable Development Goals 1; 11 and 13 would very much find pronunciation in these three elements on a parameter of development and transformation of disaster management capabilities and institutions in Zimbabwe on local–village; district, provincial and national level. Multilevel institutional disaster management capabilities development on three parameters namely prevention, mitigation and preparedness would mean a considerable reduction in loss of human life, assets, infrastructure and wildlife and industry on one hand. On the other hand, the absence of multilevel approach would mean adverse and devastating losses in all parameters with development increasingly remaining very obscure. On a parameter for comparison of disaster management capabilities and institutions within underdeveloped; developing and developed nations or alternatively low income, middle- and high-income nations where Zimbabwe categorises as an underdeveloped or low-income poor nation, considerable value would be received depending on the degree of dispensational receptiveness or non-receptiveness in Zimbabwe. More often than not, institutional development would take the form of:

- Standby competent disaster management special teams;
- available resources for emergencies on every level;
- balance of skilled and unskilled manpower in disaster management;
- advanced early warning satellite locations and communications;
- preliminary specialised evacuation holding camps for local people and their assets and domesticated animals;
- special emergency service centres; catering; life sustaining and welfare units in health and education and rehabilitation;
- specialised evacuation pool vehicles, helicopters or fleet trains and aviation;
- special coordinating unit with local and external disaster management institutions;
- Disaster management teams with capacities to fulfil local and international obligations in humanitarian crisis situations as floods, earthquakes, wildfires;
- rehabilitation and development special units comprising local and international disaster management cadres operating under specific cooperation agreements.

Systematically, disaster management capabilities comprehensively developed on every level highlighted would mean Zimbabwe is poised for greater and faster transformation in a world of competition. Judging by the magnitude of extensive loses Zimbabwe incurred for both the 2014 Tokwe Mukosi Floods and the 2019 Cyclone Idai disasters relative human loss of life; asset and infrastructure damages wildlife and domestic animals' loss, it becomes very clear how the risk management side was and still remains ill-equipped and ill-organised. In all circumstances, the risk management side and phase are quite critical for development of national disaster capabilities. They have the otherwise earmarked projects in other sectors of the economy. They should get funding for more productive and constructive engagement as implied under development cooperation programmes as Agenda Africa 2063 and the UNSDGs.

As regards the crisis management side of disaster impact that is in the aftermath of a strike, Zimbabwe and most nations in Sub Saharan Africa and the developing world have simply waited for a disaster to happen and then act on it as observed in the Editorial Comment of the *Online Sunday Mail* edition of March 17. 2019 and the subsequent huge 331.5 US$ million requested by the United Nations for Humanitarian Assistance to the Cyclone Idai rehabilitation process. Securities and guarantees obtainable from the Constitutional Acts as much as disaster management and strengthening of its capabilities is concerned have remained quite questionable even as repealing of the 1989 Civil Protection Act into a stronger Emergency Preparedness and Disaster Management Act (EPDM) since 2005 to date 2019 has remained obscure, protracted, unpredictable and hanging in the pipeline. By interpretation, this denotes a very fragmented and unstable disaster management institution and capabilities. Even as development cooperation makes provisions for consolidation of these capabilities the fact that Zimbabwe and any other sovereign nation has a signatory status to international conventions on disaster management compels her to act according to provisions. The risk management side proved overwhelmed. It inherently could not be better for Zimbabwe on the crisis management side as this sooner exposed not only the scale of unpreparedness, but above all the pronounced level of socio-economic cultural underdevelopment of Zimbabwe as a heavily indebted sovereign state in a competing world (Population Reference Bureau, 2015).

On the crisis management side, Fig. 3.1 depicts relief/response, recovery, rehabilitation and reconstruction. However, the socio-economic cultural, scientific and technological level; and degree of receptiveness of a given nation determines not only the level of preparedness to disaster but also the degree of response, recovery and rehabilitation on the strengths of supposed existing disaster management systems that are in place. With Zimbabwe being a highly indebted country, low income economy registering hardly beyond 1800US$ GNI per capita, saddled by a 12 US$ billion heavy domestic and foreign debt that is almost four decades after independence and also with a 156/178 corruption index ranking, the likelihood of development of national disaster management capabilities for speedy restoration and rehabilitation in the near future is placed in great doubt (Population Reference Bureau, 2017; Transparency International, 2016). It is among many in these very circumstances that Jose Ugaz (2016) Chair Transparency International observed that "in too many countries people are deprived of their most basic needs and go to bed hungry every night because of corruption, while the most powerful and corrupt enjoy lavish lifestyles with impunity". Zimbabwe is not spared in this observation as partly implied by the depressive macroeconomic data highlighted above on GNI per capita, debt crisis and corruption index.

Figure 3.1, illustratively locates "development" with "disaster impact" being above directly opposite implying that disaster management systems and inherent capabilities are just but one dimension within the process of national transformation as it rebounds in global development frameworks.

Development in this case presupposes adoption of comprehensive national socio-economic cultural development plans with an actively alternating inward and outward orientation for receptive nations as they maximise on cross sector value. Inherently this means disaster management capabilities of Zimbabwe and other developing nations thrive not only by themselves but in circumstances of progressive national planning within a global context where value derived is most probably maximum.

In concluding the discussion on strengthening resilience through disaster management phases and deepening partnerships with developed world regions, chances are that disaster management systems of Zimbabwe or any other nation have a greater likelihood of being strengthened to guarantee security of human life; minimum loss of life; less destruction of infrastructure and industry and wildlife through exploitation of value found in development cooperation programmes and partnering with more economically and industrially advanced nations that possess developed disaster management systems and greater experience in dealing with various types of anthropogenic and natural disasters.

Urban Resilient Infrastructure Development and Post-Disaster Transformation Programmes

Quality assurance and standardisation of domestic, industrial and commercial structures and infrastructure is very critical not only for urban but, rural settlement as well since disaster strikes anyhow. Construction of metal, wire, concrete and iron rods reinforced structures in urban settlements guarantees minimum loss in the event disaster strikes as climate change adversely impacts on all facets of human life and economy. Construction of reinforced structures should be viewed as a critical dimension in strengthening disaster management systems and capabilities of Zimbabwe. Aspiration 1 in Africa Agenda 2063 and United Nations Sustainable Development Goals 1, 11 and 13 not only accentuate development of effective national disaster management systems but also construction of a resilient infrastructure capable of withstanding the adverse impact of climate change (Africa Agenda, 2015, High Level Panel Report, 2015). Infrastructure maintenance programmes for domestic, industrial and commercial structures constitute a critical dimension on strengthening disaster management capabilities, systems and institutions in step with standards as pronounced in commonly agreed international conventions, (ibid.: 2015). With a focus on the internal and external circumstances, disaster management institutions in Zimbabwe need to have risk and crisis sides capabilities development taking advantage of opportunities availed by the development cooperation programmes through all levels from the local-village to the national level.

In the contemporary, twinning development programmes of urban settlements or cities in developing and developed nations could assist in reinforcement of standards and construction of resilient infrastructure in most urban settlements; a process in that there exists simultaneously disaster capabilities development, urban transformation and modernisation by world standards. By all logic of reasoning, this situation is poised to benefit most from the UNSDGs Goal 13 that stipulates; "Taking urgent action to combat climate change and its impacts" among other means by strengthening resilience and adaptive capacity to climate-related hazards and natural disasters in all countries. This objective is, among other factors, achieved through integrating climate change measures into national policies, strategies and planning; and by improving education, awareness-raising and human and institutional capacity on climate change mitigation, adaptation, impact reduction and early warning. Zimbabwe would strategically benefit from readily available commitment on implementing the commitment undertaken by developed-country parties to the United Nations Framework Convention on Climate Change to a goal of mobilising jointly $100 billion annually by 2020 from all sources to address the needs of developing countries in the context of meaningful mitigation actions and transparency on implementation and fully operationalising the Green Climate Fund through its capitalisation as soon as possible. This provision further promotes mechanisms in raising capacity for effective climate change-related planning and management in least developed countries, small island developing States, including focusing on women, youth, local and marginalised communities. Ultimately, acknowledgement of the fact that the United Nations Framework Convention on Climate Change is the primary international, intergovernmental forum for negotiating the global response to climate change is vital (High Level Panel Report, 2015).

For receptive nations that Zimbabwe should essentially be, mutually beneficial provisions availed under the United Nations Conventions not only consolidate disaster management systems but by all means promote industrialisation programmes that potentially lead to transformation not only of resilient urban settlements but also resilient rural and resettlement areas assuming that good planning.

It is critical to note that a full subscription to the Africa Agenda 2063 Aspiration 1 and the United Nations Sustainable Development Goals, especially Goal 13, can potentially transform not only the disaster management systems of Zimbabwe, but its whole economy as the provisions this Goal takes into account are typically core to processes on industrialisation driven national economic transformation.

Conclusion

Concluding the debate on disaster management capabilities in Zimbabwe relative recovery, loss and potential for urban transformation in a context of Africa Agenda 2063 as an appendage of the United Nations Sustainable Development Goals three key positions emerged, namely: status of local instruments on disaster management and their application by the state; potentially existing transformative international tools and abounding existing prospects on disaster management capabilities development. Accordingly, in the light of these three observations we can conclusively state that:

- The current situation with Zimbabwe disaster management systems and capabilities denotes instability and fragmentation of the system that generates incapacitation in the final analysis. Such a scenario, among other factors, arises from a very protracted and well relaxed transition from the 1989 Civil Protection Act to an envisaged stronger but obscure Emergency Preparedness and Disaster Management Act (EPDM) still in the pipeline to date (August 2019) since 2005.
- The disaster management systems and capabilities development value and opportunities availed by corresponding international conventions, such as the Hyogo Framework and Sendai Framework on Disaster Management, are not being fully taken advantage of by Zimbabwe in developing its disaster management capabilities to match global standards in development of disaster management systems and national capabilities on management of disasters.
- Novelism of the study involves developing disaster management capabilities model for Zimbabwe in step with the world. There is greater rationale resorting to cooperation in disaster management as a strategy for minimisation of losses on one hand and uplifting the affected urban settlements to ranks of modern global

cities as afforded by robust industrialisation programmes grounded in development cooperation frameworks, like Africa Agenda 2063 and the United Nations Sustainable Development Goals in a highly competitive world.
- Zimbabwe could successfully establish a highly resilient disaster management system and capabilities strategically level by level from local-village, district, provincial and national level as a guarantee for minimum losses of human life, infrastructure destruction; domestic and wildlife loss in the event disaster strikes. Indeed, great losses could be avoided through a competent and dovetailed management of the risk and crisis management sides well ahead in the alert phase.
- The backdrop of greatly transformative opportunities availed by international conventions makes the process of consolidating disaster management capabilities in Zimbabwe a unique development opportunity as Zimbabwe enjoys a signatory status to the United Nations Sustainable Development Goals, especially Goal 13 that can potentially transform not only the disaster management systems of Zimbabwe but its whole economy to the extent provisions this Goal embraces are by far typically core to processes on industrialisation driven national economic transformation.

At a closer and more critical examination, while both local and international circumstances for Zimbabwe are conducive for consolidation and development of disaster management systems and capabilities, there is pronounced apathy and procrastination on the powers that be as manifest by a still outstanding adoption of the long overdue Emergency Preparedness and Disaster Management Act (EPDM); and poorly funded and commandeered responses to disaster strikes as occurred with the 2014 Tokwe-Mukosi Floods victims and the 2019 Cyclone Idai tragedy.

References

Africa Agenda. (2015). *Agenda Africa 2063. The Africa We Want* (Final ed., Popular Version). Auc.

Christopherson, R. W. (2009). *Geosystems: An introduction to physical geography* (7th ed., Pearson International Edition, p. 687). Pearson Prentice Hall.

Emmanuel Mavhura. (2016). Disaster legislation: A critical review of the civil protection act of Zimbabwe. *Natural Hazards: Journal of the International Society for the Prevention and Mitigation of Natural Hazards, 80*(1), 605–621.

High Level Panel Report. (2015). *A new global partnership: Eradicate poverty and transform economies through sustainable development*. Unsdgs. UN Publications.

Magaya, P. W. (2017). *7 things that Africa must change* (p. 89). Yadah Publishing House.

Mudzingwa, D. (2015). *Zimbabwe: Coerced into Precarious Resettlement*. Human Rights Watch.

Pieterse, E. (2008). *City futures* (p. 206). Zed Books Uct Press.

Population Reference Bureau. (2015). *World population data sheet. With special focus on women's empowerment. Inform, empower and advance*. Prb.

Population Reference Bureau. (2017). *World population data sheet. With special focus on youth. Inform, empower and advance*. Prb.

Prevention Web. (2019). The knowledge platform of disaster risk reduction. Sendai FRAMEWORK MONITOR. https://www.preventionweb.net/sendai-framework/sendai-framework-monitor//. Accessed 09.06.2019.

Prevention Web. (2019). The knowledge platform of disaster risk reduction. Hyogo Framework for Action. https://www.preventionweb.net/sendai-framework/hyogo/. Accessed 09.06.2019.

Stutz, F. P., & Warf, B. (2012). *The world economy. Geography, business, development* (p. 435). Prentice Hall.

Tau, M., Niekerk, D., & Becker, P. (2016). An institutional model for collaborative disaster risk management in the Southern African development community (Sadc) region. *International Journal of Disaster Risk Science, 7*, 343–352. https://doi.org/10.1007/s13753-016-0110-9. Accessed 04.06.2019

The Craft Draft Constitution of the Republic of Zimbabwe. Copac Constitution Select Committee. Ensuring A People Driven Constitution. January 31.2013. Harare: Copac, p. 172.

Transparency International. (2016). Transparency International Corruption Index. Transparency International: The Global Coalition Against Corruption. www.transparency.org/cpi. Accessed 15.06.2019.

Warfield, C. (2009). Disaster management cycle. https://www.disastermanagement.org/disaster_management_monitor//272404370.html. Accessed 10.06.2019

William, K. C., & Lindsey, B. S. (2011). *World politics: Trend and transformation. 2010–2011 Edition* (p. 670). Wadsworth Cengage Learning.

Xinhua Report. (2005). Zimbabwe to Enhance Disaster Management. Comtex. July 27. Harare. *Scribbr* https://reliefweb.int/report/zimbabwe/zimbabwe-enhance-disaster-management. Accessed 13.07.2019.

Zhou, Q., Huang, W., & Zhang, Y. (2011). Identifying critical success factors in emergency management using a fuzzy dematel method. *Safety Science, 49*(2), 243–252.

Godfrey Chikowore is a Research Associate at University of South Africa (UNISA, CEMS), Senior Research Fellow and Lecturer at Midlands State University (MSU) in Zimbabwe. Author of the incisive book "Transformative essence of Japan Official Development Assistance Policy: Significance for millennium regional economic cooperation and integration in Southern Africa and Sub Saharan Africa 1980–2015 and Post 2015 Agenda and beyond". Visiting Research Fellow at IDE JETRO (Japan). Former member of Council of the University of Zimbabwe and Director of the University of Zimbabwe Institute of Development Studies.

John Nhavira is a PhD Holder and a Senior Lecturer in the University of Zimbabwe, Faculty of Commerce and Business Studies. He is specialised in Finance and lectures in Corporate Governance; Capital Markets; Banking Theory and Practice; Corporate Banking.

Terence M Mashingaidze is Executive Dean of the Faculty of Arts at the Midlands State University, Zvishavane Campus, Zimbabwe. A PhD holder he is Senior Lecturer in the Department of History. Obtained his PhD in History from USA, University of Minnesota. He has extensively published in the intersecting areas of African liberation movements and nationalism; conflict and reconciliation processes; and development induced spatial dislocations.

Constantine Munhande is a Senior Lecturer and holder of a Master's Degree with the University of Zimbabwe. He is Head of Department of Development Studies in the Faculty of Arts Midlands State University, Zvishavane Campus. He is specialised in migration, political economy and development studies.

Chapter 4
Planning and Design as Defining Parameters for Urban Resilience: The Case of Zimbabwe

Andrew Chigudu

Abstract The chapter aims to examine the impact of urban planning and design on the developing world. Urban planning and design have become increasingly important due to changes occurring in global development. This chapter seeks to explore the significance of urban planning and design in the growth and development of sustainable urban centres. Particular emphasis was given to urban centres in the developing world. Urban planning and urban design are different but closely related disciplines that strive to create sustainable towns and cities. Urbanisation has brought with it challenges that most developing countries, such as Zimbabwe are not equipped to handle. There has been a high rate of urbanisation both in terms of the increase of population in cities and the spread of development in urban areas. This has been accompanied by problems, such as overpopulation, overcrowding, shortages of resources and the growth of slum settlements. The projected 50–70% increase in the urban population by 2050 has revealed the importance of urban planning and design in catering for growing populations, especially those in Africa and Asia. Data for this chapter was collected from primary and secondary sources, such as population statistics, reports, journals and the Regional Town and Country Planning Act. Other data was collected through interviews from practicing town planners and observations from existing urban infrastructure in Zimbabwe. The data collected reveals that some countries and their cities have adapted to urbanisation and its accompanying challenges. Zimbabwe has attempted to adopt urban policies from first world countries. The adoption in Zimbabwe of some aspects of urban policies is practicable, but it is not feasible in other facets since the resources and immediate problems are not the same and there are different ways of dealing with them. It is critical for developing countries to seriously consider urban planning and design to come up with contemporary designs that are resilient to current urban challenges. Contemporary urban planning and urban design should centre on marginalised and disadvantaged areas that are constantly being left out in the planning arena.

A. Chigudu (✉)
Department of Architecture and Real Estate, University of Zimbabwe, 2024 Cardinal, Arlington, Harare, Zimbabwe

I. Chirisa and A. Chigudu (eds.), *Resilience and Sustainability in Urban Africa*, Advances in 21st Century Human Settlements,
https://doi.org/10.1007/978-981-16-3288-4_4

Introduction

Urban planning refers to the creation of maps that show that activity is undertaken in the quest to control the city's development (Dear & Scott, 2018). Planning came into use as a measure to reduce health risks in cities after the Industrial Revolution in Britain. Cities were crowded, with poor sources of drinking water, that led to the spread of diseases, such as cholera, typhoid and tuberculosis. Urban planning provided knowledge about where human waste goes and the sources from that to get drinking water. When urban planning was adopted in African countries, it was used to demarcate the residential from the commercial and industrial areas. Urban planning separate land uses that cannot be mixed, as in the case of housing developments within an industrial site (Rathore et al., 2016). Such a mix may lead to noise pollution and health hazards because of noxious industrial emissions. Urban planning therefore, provides a way of making land uses complement each other for the best possible outcome.

Urban design refers to the relationship between men and the built and natural environment and the creation and bringing to life of urban plans. It is concerned with the structure and layout of the city, not only where everything goes but how it will look, feel and function (Sennett, 2017). Urban design focuses on the future construction of strong buildings that can weather any type of storm and on infrastructure that meets construction standards and satisfies the needs and wants of the people. Issues of sustainability are addressed through conscious design and construction of urban infrastructure that include buildings, streets, public places, walkways and street furniture (Russo & Comi, 2016). Urban design is critical in the development and management of resilient urban infrastructure. Urbanisation refers to the growth in the population of people living in urban areas that is mostly caused by rural to urban migration (Hugo, 2017). It also refers to growth of development in urban areas in terms of infrastructure.

Urbanisation is beneficial because an increase in the urban population may lead to an increase in the labour force, that translates to high productivity (Cai et al., 2018). It also culminates to an increase in government revenue as more people will be paying tax. However, rapid urbanisation can lead to a boom in population, that an urban area may not be equipped to handle. A lot of people will be in the urban areas, leading to unemployment, since the labour available exceeds the workforce needed. This leads to a reduction in the GDP per capita, that leaves families without sufficient money to cater for their basic needs (Muzzini et al., 2016). There is social pressure on resources and the environment in towns and cities. Shortages of employment and money force people resort to desperate measures.

Most urban citizens cannot afford to buy houses or pay rent, and end up constructing shacks to live in. Owing to the lack of sanitary water, people just use whatever water they can find. This creates favourable conditions for the outbreak of water-borne diseases, such as cholera and typhoid, as what occurred in Zimbabwe in 2008 and 2018 respectively (Wood et al., 2017). Another consequence of rapid urbanisation is land degradation, since shacks and squatter settlements are not planned for and are constructed in a haphazard manner. They cause the ground to become compact, that reduces infiltration, and accelerates soil erosion, leaving the ground bare. Since people also cut down trees to create their settlements, deforestation occurs, resulting in erosion and the destruction of vegetation, escalating the risk of climate change (da Silva et al., 2017).

The whole city does not collapse when it is placed under strain, as some areas remain standing and adjust to the changes. This is referred to as urban resilience (Meerow et al., 2016) or the ability for persons, societies, productions and organisations to continue to grow despite shocks, whether good or bad that rock the country. Most cities have buildings or roads that cease to exist in the economic market after they have been subjected to strain. This shows their non-resilient nature and the need for them to be altered to create sustainable twenty-first century infrastructure.

Urban areas provide many functions and services. The 'pull' of these services draw people to the city (Fistola & La Rocca, 2017). As the city functions, a lot of heat tends to be produced from all the energy being used. People produce their own heat, as does the machinery used and the sun. Intense heat is produced, since high temperatures are trapped within the city, creating an urban heat island (Mirzaei, 2015). The temperature in the city, causes a lot of convectional rainfall and makes it uncomfortable to live. For this reason, various designs have been developed to reduce the urban heat island and its effects.

The thermal heat from the city has nowhere to go, so it collects in the city. Various micro-climates are designed within the city to stave off the heat, thereby, reducing the effects of the heat island (Akbari et al., 2016). These micro-climates include the placement of fountains or water sources in the CBD. Harare, for example, has fountains in the Africa Unity Square, even though they rarely work. The Eastgate shopping mall also has fountains, that are no longer in use. However, the water helps to reduce the heat felt in the city, thereby regulating the temperature in the CBD, and reducing the heat island. Green buildings are another way through that urban designs seek to reduce the effect of urban heat islands. The vegetation helps to absorb some of the carbon dioxide being produced in the city, thereby, reducing high temperatures that are trapped by the carbon blanket (Knight et al., 2016). The trees also provide shade from direct sunlight. Buildings are therefore, being designed with an element of green in them as plant growth is being encouraged, for instance, Eastgate Mall in the CBD of Harare. Development of green buildings seeks to protect the environment and the atmosphere, preventing or reducing the effects of climate change, such as droughts, floods, and erratic rainfall, shifting in weather patterns and heat waves and cyclones (Hirabayashi et al., 2013). This also influences the design of resilient cities and neighbourhoods, providing adequate services and resources to the people.

Conceptual Underpinnings

Urban planning is a broad topic, that looks at the placement of land uses, creating an urban area. The ordering of different but related land-use enable that complementary uses are well connected. Urban design then looks at the area created and makes the environment liveable. It creates places and spaces that people enjoy to live in and want to be associated with. Urban design is made up of various elements, that include legibility, walkability, lighting and safety. These elements help to ensure that an environment is adequate and flexible enough to allow for the growth and development of that area.

Legibility

Legibility looks at how clear and understandable a city is. It focuses on the pattern and order of roads and streets (Taylor, 2009) and this is usually seen from a bird's eye-view of the city. Some countries and cities have adopted various shapes and designs. Zimbabwe has a grid pattern, where the roads and streets seem straight and form a grid shape. There are parallel roads, such as Robson Manyika Road, Robert Mugabe Road, Jason Moyo, Kwame Nkrumah and Samora Machel, that run along the Harare CBD. Parallel roads. such as Angwa Street, First Street, Second Street and Fourth Street cut across other roads, creating a grid iron pattern. This makes the city legible as the pattern is the same and not so jumbled that one may get lost whilst trying to meander through the CBD. Clear patterns give the city a character that one can easily understand and allow for safety and direction.

Connectivity

The concept of connectivity relates to the linkages of residential, commercial and institutional land uses by roads (Ahern, 2013). It ensures that existing roads and streets allow for route from one place to another. This ensures the connection of various areas to the city. With connectivity, the market al.so increases, thereby improving the economy. Connectivity also ensures that there are no or very few marginalised areas in terms of services and linkages owing to the presence of roads enabling access. This improves accessibility of basic services to all urban citizens and also fosters better urban life.

Diversity

Diversity is another concept that should not be overlooked in urban design. It allows for a variety of services and commodities to be available in a particular location (Talen, 2006). Giving people a range of options in terms of products and services reduces monopoly and ensures stable prices and variety in commodity. This concept also makes an area vibrant because of the many activities being offered. The idea of diversity is more pronounced in Harare and Bulawayo in Zimbabwe.

Walkability

The city of Harare has adopted the concept of walkability. This relates to the development of flats and housing facilities within and at the periphery of the CBD, such as the Avenues flats and Calder Gardens along Sam Nujoma. This allows for people to walk short distances to work (Ewing & Handy, 2009). According to planning experts interviewed, walkability was initiated by the British colonial government during the colonial period when it developed the Mbare flats just outside the CBD so that the black inhabitants could walk to work. Such a design measure contributes to the sustainability of the environment because it reduces the carbon foot print left behind by vehicles used to ferry people in and out of the city. Areas near the city have very expensive rentals, making them prime land and the most sought-after accommodation facilities.

Mixed Use

The concept of mixed use is also being adopted in Zimbabwe, particularly in Harare, Bulawayo and Gweru. This is seen through the development and existence of various land uses within the same area (Forsyth, 2015; Montgomery, 1998). The Harare and Bulawayo CBDs, for instance, have commercial areas, institutional and health facilities. One can find a retail shop, a wholesale store, a clinic and banking facilities within it. This means that everything that a person may need is attainable within the same boundary. Mixed use helps to protect the environment, especially the ozone layer, since it also reduces use of automobile as urban citizen's park and walk for services.

Character

Building height is one of the most sensitive issues in urban design. Areas have certain building heights in accordance with the land uses in that they are located (Ahmadi et al., 2017). It is not appropriate to have a twenty-storey building in a residential area because tall buildings block the sun from reaching other areas. They cast shadows on shorter buildings, preventing natural light from being used in the building that is overshadowed. This also contributes to the character of the area, either beautifying it or creating a haphazard and unliveable environment. The notion of the character of the area was first introduced through the neighbourhood concepts, such as the

Garden city concept (LeGates & Stout, 2015). Character in urban design also relates to historical buildings. Harare has such buildings along Robert Mugabe way that cannot be demolished. This is good in the sense that it provides the city character and an edge, but it also slows down development as those buildings cannot be altered but can only have a face lift.

Adaptability

Cities need to be adaptable and able to transform themselves in accordance with changes in technological, political and economic advancements or setbacks (Sholihah & Heath, 2016). Adaptability ensures that no area is left unproductive as it would have been left derelict, turning into a 'white elephant'. Change, that is inevitable, should be embraced as it is an opportunity for growth. Adaptability promotes the sustainability and resilience of infrastructure. Most buildings in the Harare CBD have adapted to uses that are different from those for that they were initially planned. By way of example, there are now more churches, food outlets and boutiques than there were previously, that shows that the buildings were adaptable since they accommodated different uses.

Safety

Another vital concept in urban design is safety. It focuses on the secure movement and habitation of people in an environment or neighbourhood. Safety can be achieved through comprehensive planning and designing, taking into account building height, orientation and surrounding infrastructure and vegetation (Cai & Wang, 2009). Open spaces are needed, especially at the front or entrance to flats or buildings so ensure safety. From interviews conducted with town planners at City of Harare, one may consider the ample space that is in front and around the flats in Glen Norah B. It provides a clear view of who is approaching, protecting the area from intruders, as one can quickly notice the presence of someone who does not belong in that area.

Recreational Facilities

Recreational facilities are another vital element of urban design. These are designed to be located in all residential, commercial and industrial land uses and give a break of scenery from buildings to natural elements, such as trees, grass and flowers in parks (Polat & Akay, 2015). Recreational facilities also include play areas for children and accommodate all age groups. This makes them inclusive and sustainable and resilient in the face of change. Harare has the Harare Gardens and the Africa Unity Square whose trees provide shade to people who frequent them. Recreational facilities also include gyms, that allow for exercise and promotes healthy lifestyles amongst the people, reducing health risks, such as cardiac arrest.

Lighting

Lighting is a vital aspect in urban design, since it provides safety, especially during the evening. Adequate lighting is required in all areas to ensure that people can move around freely and safely (Brandi and Geissmar-Brandi, 2006). The preservation of human life is an important goal of proper lighting. There needs to be shade to control the light and prevent it from penetrating into other people's homes. The concept of lighting also refers to natural lighting, that comes from the sun and the moon. There is need to promote the use of natural light, hence, urban design accommodates large window surfaces that allow light to pass through. This concept has been adopted in Zimbabwe with new buildings and old ones are being renovated to include window walls. Most houses are now being constructed with window walls.

Research Methodology

The section seeks to assess the significance of spatial planning in Zimbabwe as a means to ensure sustainable urban development. It falls within the qualitative research paradigm where site visits and observation were undertaken as the source of the bulk of the primary data. Observations were also made of the buildings being built and refurbished within the city. Other relevant information was collected through interviews with planning experts and the analysis of published statistical and informative journals or reports. These provided information on cities and the concepts that they have adopted and also on whether the developments have proved to be resilient or just a waste of resources, money and time. Chronological analysis and comparison were also carried out to gather information about where some concepts began and why and how they have adopted.

Results

Urban planning and design have taken a backseat in today's cities, but it is vital for them to steer development once again because of the increasing population (Van Melik et al., 2007). They are important because without planning and design, cities will grow unchecked, that is risky.

Urban planning and design are very important because they give cities a blueprint as to where they are going (Kunzmann, 2016). In Zimbabwe, urban planning and design application in the real world is very limited for a variety of reasons, including the inaccessibility of funding to steer planning and design. This has left the growth and expansion of country's cities at the mercy of organic growth. After the attainment of independence, people moved to urban areas from rural areas owing to a variety of pull factors. These include the search for employment in urban areas as opposed to sitting idle in rural areas. People also wanted to be able to pay tax and so they moved to cities. There was also a desire to attain education, particularly tertiary education, that could not be accessed in almost rural areas. This led to a population boom in urban areas. The migration to cities did not stop then, but continued to take place and is still taking place, placing further strain on the cities. The population in cities also grew owing to the removal of white supremacy rule that had imposed legislation prohibiting the movement of people to cities (Dzingirai, 1998).

In the face of rapid urbanisation, urban planning and design are vital. Their absence has given rise to a number of phenomena (Patel, 1988; Potts, 2016). Rapid urbanisation has led to pressure on resources, such as water and sewer, housing and power. A lot of squatter settlements have emerged due to the shortage of housing facilities and economic challenges. Squatter settlements and illegal structures have sprouted all over the urban set up. This has led to land degradation owing to the scarring of land, the compactness of soil and soil erosion. In Zimbabwe, the existence of squatter camps had reached an alarming state where they sprouted amongst planned areas. There was need for the eradication of this squatter camps and slums that culminated in the infamous *Murambatsvina*, that saw the destruction of all shacks and illegal structures in a bid to clean the city. Operation Murambatsvina was considered inhumane, although to planners, it was a necessary evil that had no other alternative except demolition. Population growth has highlighted the inadequacy of the sewer system, especially in the absence of urban planning and design. Quite a number of sewer pipes burst because they are exceeding their carrying capacity (Juru et al., 2019).

Planning experts indicated that there are weekly pipe bursts in high density areas, such as Chitungwiza and Budiriro. The incapacity of cities to handle pressure has led to the excavation of pit latrines in urban areas as a measure of resilience. These latrines give off a very unpleasant smell, are sources of diseases, such as diarrhoea, typhoid and cholera since flies can access them easily, and also pollute the groundwater taken up by people through boreholes and wells. The sinking of boreholes and digging of wells was a measure taken by people to provide water to their own families (Muchingami et al., 2019). A lot of residential areas lack running and potable water. Not everyone can afford to sink boreholes on their stands so this has led to the development of long water queues, that places physical and mental strain on people. Wells have collapsed, and people have drowned in them and they have also been a source of diseases, like typhoid, diarrhoea and cholera since the groundwater is being contaminated and, therefore, unsafe to drink.

Cities have tried to be resilient in a number of ways, that albeit not 100% beneficial. By way of example, there is the sinking of boreholes and digging of wells to supplement council water in major cities and towns including Harare, Chitungwiza, Gweru and Bulawayo (Muchingami et al., 2019). The people now have water for domestic use and have to take care to make it potable through boiling and applying water guard treatments. Some residents in Bulawayo's Cowdrey Park high density and Harare's new suburbs, such as Glenview Extension, Stoneridge, Caledonia and Hopley have resorted to construction of septic tanks, soakaways and pit latrines for sanitation though not permitted by planning standards. Make-shift housing structures are being constructed in some peri-urban areas that include DZ Extension, Murisa in Chitungwiza and Epworth. The housing crisis, especially in Harare, has led to invasion and development of most in-fills in Glen Norah and Malbereign. Such practices have engulfed most green breathing spaces.

Electricity shortages have resulted in people resorting to other sources of energy and only a few people have resorted to using solar systems in their homes (Miller, 2015). Some people have resorted to using generators, that are guzzling fuel that not everyone can obtain due to the continuously rising fuel prices as of 2019. Some people also resorted to using gas, that is both harmful and expensive. These options may have their downsides, but they are alternative energy sources for people. People have also resorted to using wood fuel and charcoal. The use of charcoal depletes reserves and since charcoal is reproduced after 20 million years, its exploitation is unsustainable given the extent of the current electricity shortages (Nyemba et al., 2018). The use of wood fuel causes deforestation, that, in turn, promotes climatic changes that are detrimental to the people and the environment.

Discussion

Urban planning first gained importance in the face of risks during the Industrial Revolution to people's health in Britain. It was then adopted by America and other parts of the world. Urban planning in most former British colonies in Africa, such as Zimbabwe, Zambia and South Africa had segregated planning (Ramamurthy, 2017). The urban planning undertaken during the colonial and apartheid era only benefited the whites in the cities. This led to the difference between planning standards in cities and in rural areas. These differences in spatial development have been a source of many debates in African countries, because of many differences in African countries, such as the economic, political and social ideologies. This has resulted in the African states debating on how to find a common ground where spatial planning and development has the same goal. Urban planning and design are a vital part of development because they strive to provide a suitable environment in that people can live and work in, they brighten anything that people deem to be in their best interests and that should be reflected in tangible infrastructure (Dong et al., 2018). New concepts are emerging and urban planning and design have tried to keep up with changing urban designs.

African countries are mainly adopting these new concepts from Europe, Asia and America. The prominent concept that is being adopted by other African countries is the concept of sustainability. The concept of sustainability has opened up planning and infrastructure to a lot of possibilities. For instance, many buildings and houses in Zimbabwe are now built with most walls made of glass (Braham, 2015) to conserve energy that might have been used in lighting buildings electrically since the glass walls are transparent and allow natural light to pass through. Even in this 'phase' of power cuts, many office buildings are not in the dark because of the light filtering in through the glass walls.

Another urban planning and design concept are that of green buildings. This has seen the design and construction of buildings that allow spaces for trees and other vegetation to be planted in, on and around the building. A typical example of such a building in Zimbabwe is the Eastgate Mall in Harare CBD. This works to prevent cities being only concrete jungles, allowing them to have some vegetation and plants, that contribute to oxygen supply. Urban planning and design need to focus on developing resilient buildings. This provides planners ideas on how to go about certain aspects of city development. Resilient infrastructure is a product of satisfying the needs and wants of the people. This reduces or eliminates the possibility of creating 'white elephants', in that there is very little or no activity and the infrastructure costs more than the income it should be bringing in because it is lying dormant or it is not optimally occupied. The use of the European planning system needs to be adjusted to the current urban development and the future that is different from that of the eighteenth century and the economic, social, cultural and political state of the countries.

The adaptation of the European planning system has been common in Zimbabwe and Zambia. Zimbabwe, like many other African and developing countries, is following the development standards of America and Europe. The Millennium Development Goals (MDGs) provide a blueprint that countries are following to keep up with ever-changing society where they are trying to adapt to current changes (Battersby, 2017). Zimbabwe has adopted the concept of creating housing facilities near the CBD, as evidenced by flats located along Sam Nujoma, such as the Calder Gardens in Harare. This is in keeping with the principle of walkability, that reduces the distance and time spent in transit from home to work. It also reduces carbon footprint that might have been left by vehicles to and from the city, ensuring sustainability.

Countries, such as Zambia, have adopted the concept of walkability to ensure the resilience of its residential suburbs. They have done this by establishing shopping malls in different residential areas, such as in Lusaka. The CBD has been left relatively undeveloped. This reduces the risk of overcrowding in cities since everyone can access virtually anything, they need near their home without travelling the distance to the city. This also ensures sustainability, since resources are not wasted and the area is given the opportunity to be resilient. Local business people in those areas are also satisfied because they are not negatively affected by new things or locations. The country has also witnessed an increase in the number of high-rise buildings constructed. Its vertical expansion allows more people to occupy the same piece of land.

Vertical expansion reduces urban sprawl as it encroaches onto reserved land and other land uses (Zambon et al., 2019). It also increases the revenue earned because shared space in the city centre fetches more money through rents, since it is on prime land and expensive. Sustainable development is encouraged because it provides room for future generations and future developments that will not have the natural environment in harmony with the human environment. It also ensures resilience because it can withstand the test of time. Sustainable development allows for the cautious use of resources, allowing for maximum profit without depleting available resources. Governments also tend to function better with resilient developments and infrastructure (Ghosh, 2017), since this affords it the opportunity to attend to other national issues without having to worry about infrastructural development that has become obsolete.

Conclusion, Policy Options and Future Direction

Urban planning and design are significant aspects of development. To reduce stress and strain on the cities, there is need for proper planning and designing, that not only takes into account existing urban populations but on the projected increase in population. This enables cities to expand and grow, with the capacity to develop and cater for the growing population in terms of social services, such as housing, water and sewer systems and employment opportunities.

- Both urban planning and design need to take into account resilience in urban development, encouraging it, rather than impeding it. Urban planning and design account for the location, design and placement of land uses in different urban areas. It accounts for what goes where, be it in residential, commercial and or institutional zones. With the proper allocation of space, development becomes planned and coherent. It ensures the proper functioning of the area. In other words, urban planning and design provides a guide of how the city functions.
- In the era of new urban concepts, one concept in particular has stood out: urban resilience. It refers to the elasticity of urban areas, their capacity to adapt to change that are usually strenuous to the environment. Resilience has come in different forms and includes the natural change of infrastructure to fit and weather the prevailing situation at a particular time, such that developments and infrastructure are not rendered obsolete by time and circumstances.
- Urban resilience offers a new face to development because a country does not need a complete overhaul but merely a few alterations when people need and want change. Due to the dynamic nature of people, change is inevitable.
- The continuously growing urban population is also another aspect that places stress on cities and their environments. The population still needs services and so it exerts considerable pressure on resources.
- In order for the concept of urban resilience to take root and flourish, there is need for government legislation to be amended in keeping with the changing urban landscape. By-laws and statutory frameworks need to be adjusted in line with new urban development concepts. By-laws and statutory instruments are necessary because without proper direction, resilience may take a stance that might actually be unfavourable to the city in the long run.

- These development concepts should be free from political influence, that may be difficult to achieve in most African countries. Embracing resilience should not be a tool used to garner votes, for example, but should be done to provide better opportunities for the people, their growth and development of the economy.

References

Ahern, J. (2013). Urban landscape sustainability and resilience: The promise and challenges of integrating ecology with urban planning and design. *Landscape Ecology, 28*(6), 203–1212.

Ahmadi, V., Farkisch, H., Irfan, A., Surat, M., & Zain, M. F. M. (2017). A theoretical base for urban morphology: Practical way to achieve the city character. *E-Bangi, 6*(1), 30–39.

Akbari, H., Cartalis, C., Kolokotsa, D., Muscio, A., Pisello, A. L., Rossi, F., Santamouris, M., Synnefa, A., Wong, N. H., & Zinzi, M. (2016). Local climate change and urban heat island mitigation techniques-the state of the art. *Journal of Civil Engineering and Management, 22*(1), 1–16.

Battersby, J. (2017). Mdgs to sdgs-new goals, same gaps: The continued absence of urban food security in the post-2015 global development Agenda. *African Geographical Review, 36*(1), 115–129.

Braham, W. W. (2015). *Architecture and systems ecology: Thermodynamic principles of environmental building design, in three parts*. Routledge.

Brandi, U., & Geissmar-Brandi, C. (2006). *Light for cities: Lighting design for urban spaces. A handbook*. Walter De Gruyter.

Cai, K., & Wang, J. (2009). Urban design based on public safety—Discussion on safety-based urban design. *Frontiers of Architecture and Civil Engineering in China, 3*(2), 219–227.

Cai, Y., Selod, H., & Steinbuks, J. (2018). Urbanisation and land property rights. *Regional Science and Urban Economics, 70*, 246–257.

Da Silva, J. M. C., Prasad, S., & Diniz-Filho, J. A. F. (2017). The impact of deforestation, urbanisation, public investments, and agriculture on human welfare in the Brazilian Amazonia. *Land Use Policy, 65*, 135–142.

Dear, M., & Scott, A. J. (Eds.). (2018). *Urbanisation and urban planning in capitalist society (Vol. 7)*. Routledge.

Dong, L., Wang, Y., Scipioni, A., Park, H. S., & Ren, J. (2018). Recent progress on innovative urban infrastructures system towards sustainable resource management. *Resources, Conservation and Recycling, 128*, 355–359.

Dzingirai, V. (1998). Migration, local politics and CAMPFIRE.

Ewing, R., & Handy, S. (2009). Measuring the unmeasurable: Urban design qualities related to walkability. *Journal of Urban Design, 14*(1), 65–84.

Fistola, R., & La Rocca, R. A. (2017). Driving functions for urban sustainability: The double-edged nature of urban tourism. *International Journal of Sustainable Development and Planning, 12*(3), 425–434.

Forsyth, A. (2015). What is a walkable place? The walkability debate in urban design. *Urban Design International, 20*(4), 274–292.

Ghosh, R. (2017). Climate-resilient development: Linking climate adaptation and economic development. In S. Frankhauser, & T.K.J. McDermott (eds.), *The Economics of climate-resilient development*: Edward Elgar Publishing

Hirabayashi, Y., Mahendran, R., Koirala, S., Konoshima, L., Yamazaki, D., Watanabe, S., Kim, H., & Kanae, S. (2013). Global flood risk under climate change. *Nature Climate Change, 3*(9), 816–821.

Hugo, G. (2017). *New forms of urbanisation: Beyond the urban-rural dichotomy*. Routledge.

Juru, T., Kagodora, T., Tambanemoto, C., Chipendo, T., Dhliwayo, T., Mapfumo, M., Kashiri, A., Gombe, N. T., Shambira, G., Tapera, R., & Tshimanga, M. (2019). An Assessment of the availability of water sources and hygiene practices in response to the cholera outbreak in Harare City, Zimbabwe, 2018. *Public Health, 2*(1), 1–9.

Knight, T., Price, S., Bowler, D., & King, S. (2016). How effective is 'greening' of urban areas in reducing human exposure to ground-level ozone concentrations, UV exposure and the 'urban heat island effect'? A protocol to update a systematic review. *Environmental Evidence, 5*(1), 3–99.

Kunzmann, K. R. (2016). Urban planning in the North: Blueprint for the South? *Managing Urban Futures* (pp. 251–262). Routledge.

Legates, R. T., & Stout, F. (Eds.). (2015). *The city reader*. Routledge.

Meerow, S., Newell, J. P., & Stults, M. (2016). Defining urban resilience: A review. *Landscape and Urban Planning, 147*, 38–49.

Miller, N.S. (2015). Survival 101: How Zimbabweans survived hyperinflation. *The Herald* Ottawa, Canada, Thursday, April 2, 2015, p. 1.

Mirzaei, P. A. (2015). Recent challenges in modelling of urban heat island. *Sustainable Cities and Society, 19*, 200–206.

Montgomery, J. (1998). Making a city: Urbanity, vitality and urban design. *Journal of Urban Design, 3*(1), 93–116.

Muchingami, I., Chuma, C., Gumbo, M., Hlatywayo, D., & Mashingaidze, R. (2019). Approaches to groundwater exploration and resource evaluation in the crystalline basement aquifers of Zimbabwe. *Hydrogeology Journal, 27*(3), 915–928.

Muzzini, E., Eraso Puig, B., Anapolsky, S., Lonnberg, T., & Mora, V. (2016). *Urbanisation and growth.* International Bank for Reconstruction and Development/The World Bank.

Nyemba, W. R., Munanga, P., Mbohwa, C., & Chinguwa, S. (2018). Feasibility study and development of a sustainable solar thermal power plant through utilization of mine wastelands. *Procedia Manufacturing, 21*, 353–360.

Patel, D. (1988). Some issues of urbanisation and development in Zimbabwe. *Journal of Social Development in Africa, 3*(2), 17–31.

Polat, A. T., & Akay, A. (2015). Relationships between the visual preferences of urban recreation area users and various landscape design elements. *Urban Forestry and Urban Greening, 14*(3), 573–582.

Potts, D. (2016). Debates about African urbanisation, migration and economic growth: What can we learn from Zimbabwe and Zambia? *The Geographical Journal, 182*(3), 251–264.

Ramamurthy, A. (2017). *Imperial persuaders: Images of Africa and Asia in British advertising.* ManchesterUniversity Press.

Rathore, M. M., Ahmad, A., Paul, A., & Rho, S. (2016). Urban planning and building smart cities based on the internet of things using big data analytics. *Computer Networks, 101*, 63–80.

Russo, F., & Comi, A. (2016). Urban freight transport planning towards green goals: Synthetic environmental evidence from tested results. *Sustainability, 8*(4), 381.

Sennett, R. (2017). The open city. *the post-urban world* (pp. 97–106). Routledge.

Sholihah, A. B. S., & Heath, T. (2016). Traditional streetscape adaptability: Urban gentrification and endurance of business. *Environment-Behaviour Proceedings Journal, 1*(4), 132–141.

Talen, E. (2006). Design that enables diversity: The complications of a planning ideal. *Journal of Planning Literature, 20*(3), 233–249.

Taylor, N. (2009). Legibility and aesthetics in urban design. *Journal of Urban Design, 14*(2), 189–202.

Van Melik, R., Van Aalst, I., & Van Weesep, J. (2007). Fear and fantasy in the public domain: The development of secured and themed urban space. *Journal of Urban Design, 12*(1), 25–42.

Wood, C. L., Mcinturff, A., Young, H. S., Kim, D., & Lafferty, K. D. (2017). Human infectious disease burdens decrease with urbanisation but not with biodiversity. *Philosophical Transactions of the Royal Society b: Biological Sciences, 372*(1722), 1–14.

Zambon, I., Colantoni, A., & Salvati, L. (2019). Horizontal vs vertical growth: Understanding latent patterns of urban expansion in large metropolitan regions. *Science of the Total Environment, 654*, 778–785.

Andrew Chigudu is a DPhil holder and lecturer with the University of Zimbabwe. He is an academic and seasoned spatial planning practitioner. Andrew is the Executive Director for Full Life Open Arms Africa Investment since November 2019. Before this he was the Managing Director for Hello Project Developers for 8 years. He has authored and co-authored in refereed journals. Andrew is the current Vice President for Zimbabwe Institute for Regional and Urban Planning (ZIRUP).

Chapter 5
Climate-Resilient Infrastructure for Water and Energy in Greater Harare

Yvonne Munanga and Shamiso Hazel Mafuku

Abstract This chapter reviews impacts, vulnerability and adaptation of urban infrastructure to the current climate changes experienced in Zimbabwe, Harare in particular, with the intention of providing a broad overview of possible solutions to the key issues affecting human settlements. The general lessons may also be relevant for a broader set of countries that are dealing with similar environmental, demographic and institutional challenges, particularly in Sub-Saharan Africa. A mixed methods approach is adopted and the chapter is based on document review, and interviews conducted with professionals in the built environment, environmentalists, policy makers and government officials. Human settlements in both rural and urban areas in Zimbabwe are being affected by the changing climate conditions. Two of the most pressing challenges in settlements are the water stress and energy issues. Construction of climate resilient infrastructure can help ease these challenges hence, the local government has to consider this alternative as socio-economic costs of climate change to infrastructure may probably be high if no action is taken to improve adaptive capacity of human settlements. More-so, non-adaptation may lead to the damage and destruction of infrastructure, that will in-turn affect various sectors of the economy. Right policy choices are therefore, critical in ensuring that future infrastructure and even present infrastructure is climate resilient and able to reduce risks among vulnerable groups.

Introduction

Zimbabwe has experienced a number of unprecedented economic, environmental and political shocks and stresses in the past few decades and this has greatly affected both

Y. Munanga (✉) · S. H. Mafuku
Department of Architecture and Real Estate, University of Zimbabwe, P.O Box MP 167, Mt. Pleasant, Harare, Zimbabwe

S. H. Mafuku
e-mail: smafuku@eng.uz.ac.zw

I. Chirisa and A. Chigudu (eds.), *Resilience and Sustainability in Urban Africa*, Advances in 21st Century Human Settlements,
https://doi.org/10.1007/978-981-16-3288-4_5

rural and urban development. A sharp drop in Gross Domestic Product (GDP), hyper-inflation, de-industrialization, closure of industries, large scale lay-off of employees, and disruption of public service delivery, coupled with recurrent drought, floods and poor harvests, have contributed to the huge influx of people from rural to urban areas. Rapid urbanisation, poverty, food insecurity, malnutrition, and environmental degradation are serious challenges in Zimbabwe, both in rural and urban areas. Climate change has exacerbated the situation for families and heightened overall community vulnerability, and is predicted to have continuing and primarily negative effects throughout Zimbabwe (UNDP, 2015).

Human settlements in both rural and urban areas in Zimbabwe are being affected by the changing climate conditions. Narrowing down to urban areas, two of the most pressing challenges in settlements are the water stress and energy issues. Promotion of construction of climate resilient infrastructure can help ease these challenges hence, the local government has to consider this alternative as failure to adapt to the changing climate may lead to the damage and destruction of infrastructure. Corfee-Morlot and Cochran (2011) consent to this view as they affirm that the socio-economic costs of climate change to infrastructure will likely be high if no action is taken. Non-adaptations could lead to the damage and destruction of infrastructure, that will affect all sectors of the economy. Resultantly, the right policy choices are critical in ensuring that future infrastructure is able to withstand increases in climate variability and mean changes, and also able to reduce risks among vulnerable groups.

It is therefore, imperative for responsible authorities to put in place policies and measures to promote construction of resilient infrastructure that will be more adaptive to the climate changes, and at the same time harnessing the natural resources that are abundant in the continent to address the issues of water stress and energy issues (Brown et al., 2011; UNDP, 2015; Meerow et al., 2015). This can be made possible through promoting specific building materials for the climate resilient infrastructure and encouraging recycling of water at household level and water-harvesting at community level amongst other initiatives. Housing and sanitation issues are worsening by the day. More-so, there has been sprouting of many new settlements in urban areas and most of them do not have water and electricity that compromises the health and safety concerns of these settlements yet part of the town planning principles is health and safety. The primary relevance of this chapter is to review impacts, vulnerability and adaptation of urban infrastructure to the current climate changes experienced in Zimbabwe with the intention of providing a broad overview of possible solutions to the key issues affecting human settlements to policymakers, practitioners and researchers in the planning and built environment in Zimbabwe. However, it is anticipated that the general lessons may be relevant for a broader set of countries that are dealing with similar environmental, demographic and institutional challenges, particularly in sub-Saharan Africa.

Theoretical Framework

This study is grounded on the sustainability theory, forming the theoretical framework guiding this research. Sustainability is the ability of a society, ecosystem, or any such ongoing network to carry on being functional into the indefinite future whilst not being exposed to malfunctioning due to exhaustion or excessive use of key resources that that network or system relies on (Foley et al., 2003). In relation to climate-resilient infrastructure, sustainability entails the ability of infrastructure to be adaptive to current and predicted climate conditions and at the same time being functional into the indefinite future while not affecting, negatively, the environment and resources at its disposal. Urban resilience in this context refers to the ability of an urban system and all its constituent socio-ecological and socio-technical networks across temporal and spatial scales to maintain or rapidly return to desired functions in the face of a disturbance, to adapt to change and to quickly transform systems that limit current or future adaptive capacity (Meerow et al., 2015).

Infrastructure in both rural and urban areas should have the absorptive, adaptive and transformative capacity in face of the climate changes experienced. Absorptive capacity refers to the ability to reduce shocks and stresses through preventative measures whereas adaptive capacity is the ability to make proactive and informed choices on alternative livelihood strategies based on an understanding of changing conditions (Meerow et al., 2015). Adaptive capacity in most instances is limited by poverty, poor public and environmental health, weak institutions, lack of infrastructure and services, marginalisation from decision-making processes and planning procedures, gender inequality, lack of education and information, natural disasters, environmental degradation, climate-sensitive resources, and insecure tenure (UNFCCC, 2007). Transformative capacity on the other hand embraces the governance mechanisms, policies, regulations, infrastructure, community networks, formal and informal social protection mechanisms that constitute the enabling environment necessary for systematic change. It recognises the need for system-level changes that enable more lasting resilience at household and community level, in this case focusing on infrastructure.

Literature Review

Urban populations in low-and-middle income nations are said to be the most vulnerable to current climate changes and therefore, likely to be disproportionately affected by the impacts of climate change, both directly and indirectly (Satterthwaite et al., 2007; Dodman & Satterthwaite, 2008). Urban growth between 2010 and 2030 is expected to crop up in low- and middle-income nations with the highest rates of growth occurring in Africa (Johnson, 2010; UNDESA, 2011). This implies that urban areas are increasingly becoming important sites for combating climate change's adverse impacts (Romero-Lankao & Dodman, 2011). Moreover, urban areas in Africa accommodate large proportions of their populations in hazard-prone areas, including coastal settlements, flood plains and steep slopes resulting from rapid urbanisation and an increase of unplanned settlements. Consequently, climate impacts, notably the increased frequency and intensity of extreme weather events are likely to intensify the existing natural hazard burdens for vulnerable populations, particularly in new settlements, informal settlements and slums (Douglas et al., 2008; Pelling & Wiser, 2009).

More-so, as a result of climate change, disaster risk will continue to grow in many countries as more people and their assets concentrate in areas exposed to weather extremes as urban populations continue to increase. In the case of Harare, populations are ballooning by the day as people continue to migrate to the city in search of greener pastures. The urban poor are also more likely to be disproportionately affected because they face manifold vulnerabilities that include higher exposure due to living in hazard-prone areas, lack of protective infrastructure, lack of state planning and assistance as far as disaster preparedness, response and recovery is concerned, less adaptive capacity mainly because of limited assets to invest in resilience and less financial and legal protection, such as proper insurance and security of tenure (Dodman & Satterhwaite, 2008; IPCC, 2011).

Africa is said to be one of the most vulnerable regions in the world due to widespread poverty, limited coping capacity and its highly variable climate (UNFCCC, 2007; Madzwamuse, 2010). Recent reports produced by the Intergovernmental Panel on Climate Change (IPCC) (2001, 2007, 2012) conclude that Africa will experience increased water stress, escalated food insecurity and malnutrition, sea-level rise, and an rise in arid and semi-arid land as a result of greenhouse gas emissions. As a result, extreme weather events (floods, drought and tropical storms) are expected to increase in frequency and intensity across the continent, thus posing a danger to human settlements. These projections are consistent with recent climatic trends in Southern Africa, including Zimbabwe. Recently (March 2019), there was Cyclone Idai that affected Zimbabwe, Malawi and Mozambique, leaving horrific trail and disrupted livelihoods of many communities. The effects of this exposure to changes in climate are worsened by the high levels of sensitivity of the social and ecological systems in the region, and the limited capacity of civil society, private sector and government actors to respond appropriately to these emerging threats. There is need for the governments to promote construction of climate resilient infrastructure to address some of these challenges, particularly water stress and energy issues. UNFCCC (2007) urges developing countries to prioritise climate change adaptation due to their higher vulnerability.

Resilience Strategic Framework

Building the resilience of vulnerable populations entails assisting people to cope with the current changes, adapt their livelihoods, and improve governance systems and ecosystem health so that they are better able to circumvent problems in the future (UNDP, 2015). This, however, requires an integrated approach to improving absorptive, adaptive and transformative capacity and this can be done through moving towards climate-resilient infrastructure amongst many other means. However, despite growing vulnerability in human settlements, urban policy in Zimbabwe does not explicitly address climate change. Outdated master plans, local plans and building by-laws are failing to effectively regulate development and address the stresses caused by climate change. Enforcement of planning laws is also another issue, especially in Harare. Some settlements are just mushrooming whilst the Council ignores these developments. These settlements in most instances lack basic services and are vulnerable to any harsh climatic conditions. The lack of a national climate change framework has been alluded to as the primary reason as to why climate change has not been properly integrated into policy by experts (Brown et al., 2011; Makonese, 2016).

Climate-Resilient Infrastructure

Cities in developing countries are already faced with enormous backlogs in shelter, infrastructure and services, and confronted with insufficient water supply, deteriorating sanitation and environmental pollution in face of the climate changes being experienced. The huge urban populace is demanding larger proportions of water while simultaneously decreasing the ability of ecosystems to provide more regular and cleaner supplies. As a result, sustaining healthy environments in the urbanised world of the twenty-first century is presenting a major challenge for human settlements development and management for the developing nations. This calls for flexible and innovative planning, especially considering the sudden and substantial changes in climate, water and energy demand for people and their associated economic activities (Khatri & Vairavamoorthy, 2007).

If not planned for, climate change can have significant implications on infrastructure. These implications can come in the form of physical and socio-economic costs. The socio-economic costs of climate change to infrastructure will likely be high if no action is taken to mitigate them. Even though infrastructure assets have long operational lifetimes, they are sensitive not only to the existing climate at the time of their construction, but also to climate variations over the decades of their use. There is therefore, need to increase the resilience of both new and existing infrastructure. This is made possible by planning ahead on ways to manage the impacts of climate change. Moreso, this is an important part of the transition to a green economy (Spelman, 2011; Brown et al., 2011). In the case of Zimbabwe, approximately $14.2billion is needed to rehabilitate existing infrastructure (AFDB, 2011). The state of infrastructure has been worsened by climate-related hazards but some of the infrastructure deterioration is attributable to neglect, mismanagement and inefficiencies of responsible institutions.

Water and Energy

Of late most literature on energy and climate change has been focusing mainly on the potential of 'green' technology to contribute to a new low-carbon economy in a bid to protect the environment (de Gouvello et al., 2008; Never, 2011). However, statistics show that less than 10 percent of the rural population in sub-Saharan Africa has access to modern energy services, with just over 20 percent of the total population connected to electric power supply (AfDB, 2008). The urban population is also experiencing electric power shortages as most of the electricity in sub-Saharan countries is hydro-electric power. Zimbabwe of late has been experiencing long black-outs averaging 12–16 h a day as a result of low water levels in the water basins and alleged maintenance issues (ZERA, 2019). As a result, energy provision remains one of Africa's principal development challenges.

Climate change is likely to compromise energy development, especially hydropower, that represents 45 percent of electric power generation in sub-Saharan Africa (Bates et al., 2008). It is estimated that overall surface water resources will reduce significantly by the year 2080 (Brown et al., 2011). Any reduction in available surface water will lead to increased water scarcity and power shortages. This calls for climate-resilient infrastructure for water and energy in human settlements to alleviate the likely challenges to be encountered. At present, water stress is being aggravated by climate change as rainfall received has continued to dwindle over the years. In Zimbabwe, at a macro-level, impacts of climate change, particularly rainfall, adversely affect a number of socio-economic sectors that include agriculture, water, health, infrastructure, energy, gender and human settlements. The persistent drought in Zimbabwe has severely strained surface and groundwater system, contributing to the country's deteriorating water supply. Surface water is the major source of water in Zimbabwe and accounts for 90% of supply. Potential to use groundwater has not yet been realised due to limitation of the required technology in Zimbabwe. Moreover, there is also limited knowledge on how much groundwater the country has (GOZ, 2010; Nyagumbo et al., 2010; Brown et al., 2011).

There is need for infrastructure with both adaptive and transformative capacity in the face of the changing climate conditions. Adaptive capacity is, however, greatly limited by poverty, poor public and environmental health, weak institutions, lack of infrastructure and services, marginalisation from decision-making processes and planning procedures, lack of education and information and environmental degradation (UNFCCC, 2007; Meerow et al., 2015). To ensure adaptive capacity in human settlements as far as water and energy issues are concerned, there is need to promote water-harvesting at household and community level, and embrace integrated urban management water systems that have an influence on the designing of buildings and communities.

Water stress problems are not unique to Zimbabwe only, but many developing nations are being affected. Available water sources throughout the world are depleting due to climate changes and this problem is exacerbated by the rate at which populations are increasing, especially in urban settlements in developing nations. As a result, there is urgent need for planned action to manage water resources effectively for sustainable development. With mounting global change pressures, coupled with existing un-sustainability factors and risks inherent to conventional urban water management, cities in developing countries are experiencing difficulties in efficiently managing scarcer and less reliable water resources. There is need for a paradigm shift in the way of managing urban water systems to address this challenge. The paradigm shift is to be based on numerous key concepts of urban water management that incorporate interventions over the complete urban water cycle; reconsideration of the way water is used and reused; and greater application of natural systems for water and wastewater treatment (Khatri & Vairavamoorthy, 2007).

The design of water distribution systems in Harare's suburbs has been based on the assumption of continuous supply. However, in most of the developing countries, the water supply system is no longer continuous but sporadic. In Harare, water supply is worsening by the day; most suburbs are receiving water once or twice a week whereas some are going for months without running water. Climate change is alleged to be the chief cause of significant changes in precipitation and temperature patterns that are greatly affecting the availability of water in cities. Additionally, population growth and urbanisation are enforcing rapid changes, leading to a dramatic increase in high-quality water consumption. Resultantly, this demand for water cannot be satisfied by locally available water resources as they are not adequate.

Nyagumbo et al. (2010) affirm the need for an integrated approach to urban water management. This involves managing storm water, waste water and fresh water as links within the resource management structure in an urban area (also termed the unit of management). Various aspects of water management, such as environmental, economic, technical and political are encompassed in this approach and the social impacts and implications. The integrated urban water management approach makes it possible to satisfy water related needs of a community at the lowest cost to society whilst minimising environmental and social impacts (Nyagumbo et al., 2010; Studer & Liniger, 2013). This is made possible by the fact that managing water use and re-use in a smaller community is easier than managing the whole town and city. More-so, communities differ in characteristics, so by managing a smaller unit; it can be possible to manage water resources efficiently. This approach can be adopted in both new and old settlements in Harare.

Furthermore, balancing the demands for water between various sectors in urban settlements now needs to be accompanied by the use of new and alternative resources, by increased recycling of wastewater that will ensure better access to safe water, reduced vulnerability to extremes and increased adaptive capacity. This can be done both at household and community level. A permutation of end-use efficiency, system efficiency, storage innovations and reuse strategies will go a long way in reducing water demand, thus boosting the adaptive capacity of communities on water stress. Water can be used several times by cascading it from higher to lower-quality needs and by reclamation treatment for return to the supply side of the infrastructure. In most of the developing countries, effective water demand management and reuse of the supplied water may be a sustainable way to reduce water stress (Khatri & Vairavamoorthy, 2007; Nyagumbo et al., 2010; Studer & Liniger, 2013). Harare suburbs can adopt this integrated urban water management approach to boost both the adaptive and transformative capacity of its settlements. The new developments can be designed with infrastructure that caters for recycling and harvesting of water.

Water-harvesting can be done at household or community level. The most common water-harvesting technique for domestic consumption is rooftop water-harvesting (Gur, 2013). Residents can easily be trained to build rain water-harvesting systems to encourage participation, ownership and sustainability at community level (Hatum & Worm, 2006). Additionally, by-laws can be updated to make it law to construct houses with the capacity to harvest rain water as this will reduce demand on council piped-water that is already in short supply, at the same time addressing the water stress in both rural and urban settlements. The cost of rain water catchment systems is covered in the initial construction costs of the house hence it is very affordable. Collected rain water can supplement other water sources when they become scarce or are of low quality.

On the energy sector, Zimbabwe has been faced with energy challenges since the late 2000s, that saw massive power outages around the country of up to 16 h a day. Zimbabwe relies on a carbon intensive model to generate electricity. About 43% of the country's electricity supply comes from coal, while 57% comes from hydropower systems. However, there is a lack of information on other sources of renewable energy in the country and existing frameworks to support the required technologies. Zimbabwe's electricity sector is dominated by the Zimbabwe Electricity Supply Authority (ZESA holdings), that is a state-owned enterprise. ZESA holdings is responsible for generating, transmitting and distributing electricity to all sectors of the economy in the country. However, since the late 1990s, ZESA holdings has failed to produce enough electricity to meet the increasing demand. The growing demand has been largely due to rising urbanisation and inadequate investment in additional and new forms of energy (Makonese, 2016).

Statistics in 2016 reveal that only 40% of households in Zimbabwe have access to electricity and this is not in line with the government's 2030 vision of providing sustainable energy solutions to all citizens, irrespective of their geographical location. To meet the rising demand, government has indicated that it will make it mandatory for all new buildings to use solar energy through regulation (Makonese, 2016; Mzezewa & Murove, 2017). Development and promotion of renewable energy technologies can provide a solution to the electricity supply and the carbon intensive economy of Zimbabwe power generation industry. Zimbabwe has a potential to be one of the leaders in Southern Africa in the development and promotion of renewable energy technologies. Renewable energy comes in the form of hydro, solar, geothermal, wind, and biomass.

Solar energy is one alternative that the Zimbabwean government can utilise to improve the country's energy generation mix. For the domestic sector, the potential for renewable energy from solar PV and solar water heaters is enormous. Though, to date, this potential has not been sufficiently exploited. One of the benefits of solar energy is that excess power generated from household solar power can be fed into the national grid. This accentuates the need for the Zimbabwean government to develop polices around renewable energy feed in tariffs. As efforts to improve power supply, recently the government, through its regulatory authority Zimbabwe Energy Regulatory Authority (ZERA), drafted an independent power producer (IPP) framework meant to incentivise and stimulate investment in the renewable sector. The authority has also come up with a renewable energy feed in tariff programme, that is still waiting government approval (Makonese, 2016; Mzezewa & Murove, 2017).

Buildings

Buildings are long-lived assets and therefore, present great challenges for adaptation to projected and unexpected climate impacts. On the other hand, these buildings also present a significant amount of opportunity for climate mitigation if well planned and designed. Climate resilient design features make homes resilient to climate vulnerabilities, such that they maintain an acceptable level of functioning and structure. Impacts of climate change unquestionably have consequences for building design and fabric, for the health and well-being of occupants. Identifying the nature and extent of buildings' vulnerabilities and resilience is essential if they are to be adaptive in a timely and effective manner (Adger et al., 2009; Alfraidi & Boussabaine, 2015; Kosanovic et al., 2018a, 2018b). In most developed countries, such as Australia, America, Italy, Japan and New Zealand, there are many assessment tools and methods being developed and used for identifying vulnerability and adaptive capacity of various systems from planning to urban governance that some developing nations are drawing lessons from. However, there appears to be a knowledge and policy gap in the area of assessment of buildings, in particular, measuring their vulnerability and aiding in creating climate-adaptable and resilient structures (Adger et al., 2009; Varshney & Graham, 2012).

The need to reduce the vulnerability of buildings to climate impacts is accelerating hence, there is need for policy makers and those able to effect change and implement adaptive strategies (Dessai et al., 2004; Adger et al., 2009; Ash, 2010). Adaptive measures can be prioritised based on observed and projected impacts. A list of criteria to aid in identifying key vulnerabilities may include but not limited to timing, persistence, reversibility, certainty, importance and equity issues of the impacts. Climate impacts have consequences on building design and retrofitting, structural durability, building techniques, selection of materials and finishes, and for operation and ongoing maintenance of buildings. Apart from this, there are also consequences for people in and around buildings as far as their health and well-being is concerned. It is therefore crucial to identify ways in that buildings may need to adapt in response to these impacts and vulnerabilities in different spatial and temporal contexts (Larsen, 2011; Alfraidi & Boussabaine, 2015).

More-so, when designing buildings in the face of climate change, it is crucial to come up with designs that have structure resilience. Structure resilience entails capability of a building's structural system, materials and foundations to continue to function in the face of internal and external change of climate and to degrade gracefully when it must. Buildings should use materials and construction methods that are durable when confronted with severe weather events (Alfraidi & Boussabaine, 2015). In Australia, they came up with a concept for climate adaptive buildings that they termed Climate Adaptive Building Shells (CABS). This concept is closely linked with terms, such as active, dynamic, kinetic, intelligent, responsive and smart buildings. CABS concept expressly focuses on the building envelope and explores the possibilities of various forms of adaptability in this shell in response to changing climatological boundary conditions and user's preferences.

This concept also ensures construction of buildings that have the ability to repeatedly and reversibly change some of their functions, features or behaviour over time in response to changing performance requirements and variable boundary conditions. Resultantly, the building shell improves overall building performance in terms of primary energy consumption while maintaining acceptable indoor environmental quality (Loonen, 2010; Dave et al., 2012). In sub-Saharan African countries, there has been increased temperatures that have resulted in a decrease in rainfall. These conditions can be taken into consideration on designing buildings. Any openings on the buildings can be used to harness the abundant sunlight through use of windows and doors that have the capacity to convert this sun into solar energy. Recently (2019) in Zimbabwe, there has been an effort by some entrepreneurs to manufacture glass for windows and doors that has the same capacity as solar panels. With the necessary technology, this can be embraced to come up with climate adaptive building shells.

Methodology

The study adopts a mixed methods approach where both quantitative and qualitative research methods are employed, complementarily, to generate valid and reliable data that respond to the research objectives with minimum bias (Collis & Hussy, 2003; Saunders et al., 2009; Creswell, 2014). The data used included both primary and secondary. Primary data was collected through interviews conducted with professionals in the built environment, environmentalists, policy makers and government officials. Secondary data was collected through literature and document review. Data analysis was through content analysis.

Results

Improving the climate-readiness of existing and new infrastructure is fundamental to tackling the impacts of climate change (Dave et al., 2012). Human settlements in Zimbabwe are being affected by climate changes and the most critical issues are that of water and energy.[1] Harare is facing critical water shortages and power shortages. Zimbabwe relies on run-off water and this sees urban areas getting their water supplies from lakes and dams. A greater percentage of electricity used in Zimbabwe is also hydro-electric power and in the face of the erratic rainfall patterns the power supply is a nightmare. Most new settlements in Harare do not have piped-water supply and electricity. As for the old or existing suburbs, they are experiencing erratic water supplies as most suburbs receive water once or twice a week and some have gone for months without water.[2] Whilst these challenges can be attributed to maintenance and management issues or inefficiencies by the Council and Power Supply Company, climate change has also played a part in this crisis.

An agreed perspective by most interviewees was that new settlements should have an alternative water supply. This can come from water-harvesting at both household and community level. Of late, some new developments, especially in gated-communities, have been reported as borehole systems have been put in place to supply the community and this can also be implemented in already existing and future settlements. This can be a better way to preserve groundwater as opposed to having every household drilling a borehole or well.[3] A good example of such settlements include the Borrowdale Villa Project that has a borehole supplying the complex with piped-water, the Stone Ridge housing development by the National Building Society whereby, two boreholes were drilled at key points to serve the whole community and a centralised septic tank was constructed, the aim being to generate bio-gas in future, thus addressing both the water and energy issues. The project is nearing completion.

On promoting water-harvesting, it was noted that reviewing the current building by-laws can be one way. The laws will ensure construction of buildings with the capacity of harvesting water. This would go a long way in reducing pressure on the already scarce water at the same time increasing the adaptive capacity of settlements to the changing climate. Water collected can be preserved for future use in the dry seasons. The communities, however, need to be educated on how to preserve the harvested water (Gur & Spuhler, 2019).

On improving the adaptive and transformative capacity of settlement on energy issues, the Green Building Council of Zimbabwe is working on means to promote green construction in the country. This will go a long way in improving the energy situation in the nation. There is need for Zimbabwe to adopt clean energy sources that will facilitate a green economy in the energy sector and at the same time enhance socio-economic and sustainable development, with minimal environmental impacts. However, there have been barriers affecting the adoption of clean energy sources, that is, development and use of renewable energy technologies. These include technical, technology issues, policy issues, economic, institutional and socio-cultural issues. There is lack of consistent policies and regulatory frameworks to support renewable energy, such as solar, bio-gas in Zimbabwe. This lack of policy affects dissemination

[1] Interview with Local Government Official.

[2] Interview with Harare City Council Official.

[3] Interview with ZINWA official.

of renewable technologies in the country. To address this anomaly, the Zimbabwe Energy Regulatory Authority (ZERA) through their parent ministry (Ministry of Energy and Power Development) is in the process of developing a renewable energy policy.[4]This policy seeks to attend to gaps in the existing energy policy. Areas of interest include incentives for increased uptake and investment in renewable energy technologies. A renewable energy feed-in tarrif (REFIT) framework has already been developed buy still to be implemented (Mzezewa & Murove, 2017; ZERA, 2019).

Economic barriers are also greatly hindering the energy sector thus, affecting the transformative capacity of communities from traditional energy sources to clean, renewable energy sources. To begin with, the diffusion of energy technologies faces financial barriers that come in form of high initial capital cost, high investment costs, high transaction costs, lack of access to capital (for businesses) and lack of disposable income (for households). SI 147 of 2010 tried to address some of these issues but was not successful. It provides for 0% customs duty on solar panels, solar lights, solar geysers and energy saving bulbs. However, this 0% customs duty only applies when one procures goods not exceeding US$200.00[5] that is not enough to power a standard core house. There is need for upward review of the cap provided by Zimbabwe Revenue Authority (ZIMRA) on importation of solar products as there is limited production of solar products locally. There is also need for special incentives through duty waivers on solar equipment. Recently; (17 July 2019) the Minister of Information, Publicity and Broadcasting Services announced on national television that cabinet had resolved that it shall be mandatory for all new construction projects to be solar powered. If this is to be successful, financial barriers have to be dealt with and this has to be made policy as well.

Moreover, there is a great need to create a vibrant renewable energy market in Zimbabwe to improve the transformative capacity of settlements from the traditional energy sources to use of solar energy. Government should be able to raise public funds to support the renewable energy market in the country. It could also put in place government procurement policies, through ZERA and also subsidise some of the products. This would promote a vibrant and sustainable development of renewable energy in the country (Makonese, 2016; Mzezewa & Murove, 2017). With the current economic situation in Zimbabwe, most of the clean energy products are inaccessible to the majority as people are struggling for basics, such as food, education, health.

To improve resilience of communities on climate change issues, one cannot ignore the fact that in Zimbabwe, there is need to review our planning standards regularly so that development moves with the current changes.[6] Planning standards guide the development of human settlements looking on all aspects of development, such as water, energy, roads and sewer, among other things. If these are updated regularly, then settlements will be planned with an inherent capacity to adapt, absorb and transform as needed with regards to the changing climate.

---

[4] Interview with ZERA officials.

[5] Interview with a ZIMRA officer.

[6] Interview with a Town Planner.

Discussion

Lack of consistent, up to date and appropriate information and knowledge on climate change issues is one of the barriers to effective adaptation in Zimbabwe. It is very difficult to plan when there is insufficient information and this compromises the resilience capacity of communities. Furthermore, there hasn't been much research on climate-resilient infrastructure in Zimbabwe (Brown et al., 2011; Mzezewa & Murove, 2017). This then affects policies in the area of planning for climate change, hence whilst there may be assessment tools and methods for identifying adaptive capacity of various systems from planning to urban governance, this knowledge and policy gap compromises efforts of assessing both existing and new buildings and settlements' vulnerability to climate impacts. Clarity and consistency in measuring and assessing settlements and buildings' climate adaptability and resilience, or the lack of it is very essential as this affects policy and planning.

In Zimbabwe, there really is a pressing call for advanced research to develop common metric and consistent methodologies of assisting how adaptable and resilient our existing and new settlements are to projected and unexpected climate impacts. This will give a clear picture of the challenges the nation faces as far as improving resilience of settlements is concerned. Moreover, this will enable the responsible authorities to come up with various priorities for adaptation. However, it has to be noted that there will be implications (both positive and negative) of the various priorities for adaptation on assessment criteria for both settlements and buildings that need to be identified if the authorities are to develop appropriate, timely and effective design and policy measures for adaptation (Larsen et al., 2011; Alfraidi & Boussabaine, 2015; Kasanovic et al., 2018).

Currently, in Zimbabwe and many other sub-Saharan nations, there exist many possible barriers to effective adaptation to climate change. These barriers include lack of knowledge, information asymmetries, lack of skills, finance issues, split incentives, and behavioural issues amongst many issues. There is need for the nations to find ways of addressing these barriers as there are significant risks in not adapting. Changes in climate over the past years has affected rainfall patterns, temperature levels, quality of air, quality of water and many other aspects. This, in turn, has posed water stress and energy shortages in many nations, Zimbabwe included. Adequate provision of urban water supply and sanitation has been greatly affected by climate change and urbanisation and is likely to become more difficult in the future if nothing is done as far as providing climate-resilient infrastructure in the affected settlement is concerned. There is need to develop appropriate technical and institutional responses to climate change as it radically changes the way in that urban water systems are managed. Interventions must be considered over the entire urban water cycle, recognising interactions between the various components of the urban water system. There must also be a new way water is used and reused in communities. Moreso, greater use of natural systems for treatment of water should be considered. The objective must be to develop urban water systems that are more robust and resilient against these uncertain future pressures (Nyagumbo et al., 2010; Studer & Lininger, 2013).

For the energy sector, the strategy adopted in Zimbabwe is to encourage the development of renewable energy. However, there is still a long way to go as the sector also has challenges on policy inconsistencies amongst other issues. The government of Zimbabwe has developed a national climate change response strategy to guide national response measures in addressing the impacts of climate change. Lack of implementation is, however, a great challenge as most strategies remain blue-prints over years. The national climate change response strategy aims to introduce policies and regulatory frameworks for renewable energy, energy conservation and energy efficiency; strengthen energy planning, research and development; and to promote low carbon energy provision and use (Makonese, 2016). Currently, it can be seen that there are a lot of initiatives taking place at the same time, that are aimed at the development of renewable energy in the country. However, it is important that all these efforts be effectively synchronised and rationalised to bring synergy between them and produce a properly coordinated result (Mzezewa & Murove, 2017). There is need for respective sectors to work together in crafting policies and coming up with ways to improve resilience capacity of communities.

Conclusion and Recommendations

It is a fact that the occurrence of frequent shifts in weather conditions and climate brings numerous direct and indirect consequences for the built environment, increases the possibility for disaster occurrence, and accordingly sets new challenges for human settlements. This calls for planning for such scenarios through provision of climate-resilient infrastructure. However, currently in Zimbabwe there is no common legislation or strategic frameworks in place that guide climate adaptation responses at national level. Development of a national level building adaptability and resilience assessment system that can be used for both existing and new settlements by policy-makers, regulatory authorities, property insurers, building design and construction industry professionals and householders could be highly beneficial in improving the adaptive capacity of settlements to the changing climate.

This chapter concludes that climate change is real therefore, it should be planned for. It is vital that both existing infrastructure and current construction in Harare and beyond have adaptive capacity to the changing climate to ensure safe and healthy environments for communities. The Ministry of Local Government, Public Works and National Housing through the urban local authorities and relevant departments should put in place policies and measures to ensure construction of climate resilient infrastructure that will be more adaptive to the climate changes at the same time harnessing the natural resources that are abundant in the continent to address the issues of water stress and energy issues. From information gathered in this research, it is evident that there are significant opportunities available for policy and decision-makers with the ongoing development of new tools, techniques and technologies for assessment, analysis and decision-making for climate adaptation. The government simply needs to invest in these new technologies that can be borrowed from other developed nations. More-so, most people are beginning to appreciate the impact of climate change and this will have a positive effect in policy formulation and adoption. In Zimbabwe, existing building regulations need a holistic review in the context of climate change adaptation; they have an important role to play in preparing settlements for a safe and sustainable future. An appropriate policy mix will be required to deliver climate adapted and resilient buildings to cope with these impacts. Additionally, existing buildings can be made climate resilient by ensuring that adaptation becomes an integral part of the housing stock upgrade process (Hildebrand, 2018; Kosanovic et al., 2018a, 2018b).

The paper recommends that;

- There is need to revise or update the current model building by-laws that set standards for construction of urban settlements. The standards should now ensure use of building materials that are climate-resilient. The designs should also be climate-resilient, for example, designing buildings with a green-building concept in mind for both residential and commercial buildings, designing buildings with the capacity to harvest water, designing buildings that harness the natural resources available in the respective area.
- Standards should also be set for construction of human settlements in rural areas because climate change is not only affecting the urban populace but the nation as a whole. The recent destructions caused by Cyclone Idai should be a lesson to the nation of Zimbabwe that rural settlements are just as important as urban settlements. The Rural Councils should come up with local plans for the rural settlements.
- As far as energy issues are concerned, the government through the relevant departments should either subsidise products for green energy or introduce tax shelters for suppliers of products for green energy to promote use of green energy, such as solar and gas in place of electricity, especially considering the current power shortages facing the nation. Once use of green energy has been embraced, Zimbabwe can also learn from other developed nations, such as Australia who have come up with a way of harnessing excess solar power from homes in the same neighbourhood and feed the power in the national or local grid. Use of electricity can then be restricted to other functions whereas gas can be used for some, such as cooking, heating, refrigeration.
- It is high time gas pipe-lines are included in layout designs for settlements, just like water and sewer. This will go a long way in reducing the demand on hydroelectric power. The issue of cost will, however, be of great concern, and this is where the government should come into play and subsidise some of the costs, especially on the initial cost outlay of plant and equipment required.
- Infrastructure resilience should be incorporated into the design of a building. Data from site analysis, together with information on the landscape and site climate, should be fed into the design process to evolve a resilient design aimed at addressing the risks of climate change. For example, the call by Cabinet to make it mandatory that all new developments be solar powered. This is a good move but it needs to be supported by a policy. If it becomes policy, the buildings will

be designed in a way that promotes utilisation of solar energy. More-so, there is need to either subsidise or offer tax incentives so that solar becomes affordable to the general populace.
- As far as the water stress is concerned, water-harvesting should be encouraged and made mandatory, both at household and community level. This will improve the current pathetic water situation in greater Harare.
- It is imperative that existing local and master plans be updated to incorporate climate changes that are taking place. This will go a long way in improving the adaptive and transformative capacity of settlements that are key aspects of resilience.

References

Alfraidi, Y., & Boussabaine, A. H. (2015). *Design resilient building strategies in face of climate change.* Waset Publiction.

Brown, D., Chanakira, R. R., Chatiza, K., Dhliwayo, M., Dodman, D., Masiiwa, M., Muchadenyika, D., Mugabe, P., & Zvigadza, S. (2011). *Climate change impacts, vulnerability and adaptation in Zimbabwe*, IIED Climate Change Working Paper No. 3.

Dave, M., Varshney, A., & Graham, P. (2012). *Assessing the climate change and adaptability of buildings.* City Futures Research Centre.

Gur, E., & Spuhler, D. (2019). *Rainwater-harvesting (Urban).* Seecon International.

Khatri, K. B., & Vairavamoorthy, K. (2007). *Challenges for urban water supply and sanitation in the developing countries.* Delft Institute for Water Education.

Kosanovic, S., Klein, T., Konstantinou, T., Radivojevic, A., & Hilderbrand, L. (2018). *Sustainable and resilient building design approaches, methods and tools.* PHIP Press.

Kosanovic, S., Folic, B., & Radivojevic, A. (2018). *Approach to design for resilience to climate change.* University of Pristina.

Larsen, L., Rajkovich, N., Leighton, C., Mc Coy, K., Calhoun, K., Mallen, E., Bush, K., & Enriquez, J. (2011). *Green building and climate resilience: Understanding impacts and preparing for changing conditions.* University of Michigan.

Makonese, T. (2016). *Renewable energy in Zimbabwe.* University of Johannesburg.

Meerow, S., Newell, J. P., & Stults, M. (2015). *Defining urban resilience: A review.* University of Michigan.

Mzezewa, C. T., & Murove, C. S. (2017). *Renewable energy market entry study report—Zimbabwe.* Netherlands Enterprise Agency.

Loonen, R. (2010). *Climate adaptive building shells: What can we simulate?* Eindhoven University of Technology.

Spelman, C. (2011). *Climate resilient infrastructure: Preparing for a changing climate.* Crown Copyright.

Studer, R. M., & Liniger, H. (2013). *Water-harvesting guidelines to good practice.* University of Bern.

Yvonne Munanga is a lecturer at the University of Zimbabwe. Her research interests are in rural and urban land-use change, real estate development, urban transformation and sustainability, urban renewal and regeneration, urban resilience, infrastructure and services planning and Construction Planning and Management.

Shamiso Hazel Mafuku is a lecturer in the Department of Architecture and Real Estate and a DPhil Student at the University of Zimbabwe. She holds a Master of Science Degree in Construction Project Management from the National University of Science and Technology (NUST) and a Bachelor of Science Honours Degree in Rural and Urban Planning from the University of Zimbabwe (UZ). Her research interests are peri-urban environments and sustainability, infrastructure and services planning, construction planning and management and urban regeneration and renewal.

Chapter 6
Building Urban Resilience in the Post-2015 Development Agenda: A Case Study of Harare, Zimbabwe

Elmond Bandauko

> *A resilient city must have strong infrastructure, policy and human resource response capacities to avert potential impacts of natural hazards.* (Prasad et al. 2009)

Abstract This chapter examines urban resilience building efforts in Harare. The analysis is placed within the urban resilience framework. The Post-2015 development agenda is committed to 'make cities and human settlements inclusive, safe, resilient and sustainable' (SDG 11). For this chapter, urban resilience means the ability of a system, entity, community, or person to adapt to a variety of changing conditions and to withstand shocks while still maintaining its essential functions. The four (4) dimensions of urban resilience namely infrastructure, social, economic and institutional resilience are considered. Harare was purposefully selected as it is one of the pilot local authorities under the "Partnership for Building Urban Resilience in Zimbabwe" programme by the UNDP, UNICEF and Ministry of Local Government, whose goal is to improve urban resilience and strengthen the provision of basic social services and Local Economic Development (LED) targeting unemployed youths, women, and vulnerable groups in urban and peri-urban areas (UNDP Urban Resilience Building Programme Document, 2019). The chapter concludes that there are major gaps in urban resilience building in Harare. The City is characterised by under-investment in critical infrastructure, weak urban planning and governance frameworks (including outdated policy frameworks) and lack of climate adaption planning. These factors not only work against urban resilience building, but they also hinder progress towards achieving resilient, inclusive and sustainable urban communities. For effective urban resilience building, Harare needs to prioritise investment in resilient urban infrastructure (water, sanitation, and storm water), research on the vulnerability of cities

E. Bandauko (✉)
Department of Geography and Environment, The University of Western Ontario, London, ON N6A5C2, Canada

Center for Urban Policy and Local Governance, The University of Western Ontario, London, Ontario N6A 3K7, Canada

I. Chirisa and A. Chigudu (eds.), *Resilience and Sustainability in Urban Africa*, Advances in 21st Century Human Settlements,
https://doi.org/10.1007/978-981-16-3288-4_6

and towns, internalising global and national frameworks on climate change through climate adaption planning and strengthening urban planning and governance.

Introduction

Cities are vulnerable to shocks and stresses that can erode and compromise their structures and, in turn, their resilience (EU, 2016). Shocks refer to acute and sudden events of high impact to a city's structures, such as an earthquake, fire or flooding. Stresses refer to continuous processes that erode the capacity of city's community and structures to recover properly. A resilient city has the capacities in place to shift into a different state in the aftermath of a shock or disaster while restoring its functions and services. An 'unresilient city' has limited or restricted capacity to recover, and "has high poverty and crime rates and devastated natural environment, or 'a ghost town'" (Pickett et al., 2013: 2).

The issue of resilient cities comes to the fore on the backdrop of two important facts. First, the global population living in cities crossed the 50% mark in 2008. In this regard, sub-Saharan Africa is the least urbanised but rapidly urbanising region. In Africa, urbanisation is characterised by features, such as"… unregulated growth, limited opportunities for gainful employment in the formal economy, severe environmental degradation, lack of decent and affordable housing, failing and neglected infrastructure, absence of basic social services, pauperisation, criminality, negligent city-management, and increasing inequalities" (Murray & Myers, 2006: 1). The current level of urbanisation in Zimbabwe is estimated to be 38.25% (UN-Habitat, 2010). By 2030, more than half of the country's population will be living in cities (UN-Habitat, 2010). This simply means more demand for municipal services, expansion of infrastructure and housing and growth of cities and urban economic development. The features of urbanisation point to the essence of developing sustainable, resilient, safe and inclusive cities. Second, for the first time, cities have been included in mainstream international development goals. In particular, SDG 11 focuses on making 'cities and human settlements inclusive, safe, resilient and sustainable'. SDG 11 essentially means that governments and international development agencies should prioritise policies and programmes aimed at promoting sustainability, inclusivity, safety and resilience in cities (RIPS, 2017; UN-Habitat, 2016, 2018). The two points above provide the inspiration to this chapter. The chapter seeks to stimulate debate on urban resilience building in Zimbabwean cities. It interrogates the operationalisation of urban resilience, using Harare as a case study. The chapter is based on a review and analysis of secondary data sources to understand the context of urbanisation, urban development challenges and resilience building efforts within the context of Harare.

Theoretical Perspectives

This study falls within the framework of urban resilience. The Post-2015 development agenda is committed to 'make cities and human settlements inclusive, safe, resilient and sustainable' (SDG 11). The New Urban Agenda (NUA) also commits to "strengthen the resilience of cities and human settlements, through the development of quality infrastructure and spatial planning by adopting and implementing integrated, age-and gender-responsive policies and plans, reducing vulnerabilities, enabling households, communities, institutions and services to prepare for, respond to, adapt to, and rapidly recover from the effects of hazards, including shocks or latent stresses (UN, 2016).

Urban resilience can be defined as the ability of a system, entity, community, or person to adapt to a variety of changing conditions and to withstand shocks while still maintaining its essential functions (EU, 2016; World Bank, 2015). Urban resilience is dynamic and offers multiple pathways to resilience (e.g. persistence, transition, and transformation) (Meerow et al., 2016). UN-Habitat (2018) characterises a resilient city as one that assesses, plans and acts to prepare and respond to hazards to protect and enhance people's lives, secure development gains, foster an environment for investment, and drive positive change. A resilient city is persistent, adaptable and inclusive (ibid.). The World Bank (2015) stated that a resilient city is one that has developed capacities to help absorb future shocks and stresses to its social, economic, and technical systems and infrastructures to still be able to maintain essentially the same functions, structures, systems, and identity. Urban resilience matters, especially for the urban poor. The urban poor are particularly faced with risks to their lives, health and livelihoods (World Bank, 2015).

Urban resilience has many dimensions, that include (i) infrastructure resilience, (ii) institutional resilience, (iii) social resilience and iv) economic resilience (EU, 2016; Word Bank, 2013). *Infrastructural resilience* refers to a reduction in the vulnerability of built structures, such as buildings and transportation systems. It also refers to sheltering capacity, health care facilities, and the vulnerability of buildings to hazards. The resilience of urban infrastructure and services is critically important for emergency response and the quick recovery of a community and its economy (World Bank, 2013). Institutional resilience refers to the systems, governmental and non-governmental, that administer a community. Economic resilience refers to a community's economic diversity in such areas as employment, number of businesses, and their ability to function after a disaster. Social resilience is the capacity to foster, engage in, and sustain positive relationships and to endure and recover from life stressors (ibid.).

One tool used to achieve urban resilience is planning. Urban planning has a significant impact on improving both short-term and long-term resilience capabilities of urban communities, and contributing to development gains (UNICEF, 2017; World Vision, 2016). For example, developing safely managed water and sanitation services can enhance children's access to basic services in an urban context (Plan International, 2016; UNICEF, 2018). From an economic resilience perspective, stable livelihoods for families and youth that are supported by overall economic security can help to reduce the number of children exposed to work-related risks (Plan International, 2016). If well planned, resilience-focused urban planning should inclusive and considerate of the needs of the most vulnerable and marginalised urban groups, including disadvantaged urban children and youth (World Vision, 2016).

Why Urban Resilience?

Scholars, policy-makers and development organisations have justified investing in urban resilience for several reasons. The World Bank (2015) stated that urban resilience is a critical element of sustainable development. Investing in resilience contributes to long-term sustainability by ensuring current development gains are safeguarded for future generations (ibid.). Shocks impact all aspects of development and are felt directly through the loss of lives, livelihoods, and infrastructure, and indirectly through the diversion of funds from development to emergency relief and reconstruction (World Bank, 2014a).

Investing in resilience will reduce and help prevent the impact of shocks and stresses to the city's people, physical environment, and economy. It will accelerate disaster recovery and improve the quality of life for the city's residents. Addressing urban challenges in an integrated and holistic way will help the city realise multiple benefits across sectors and stakeholders, in particular, for the poorest and most vulnerable groups. The disproportionate impact of urban shocks and stresses on a city's low-income population and informal settlements is apparent. A growing literature is drawing attention to the lack of resilience amongst the urban poor. Poor people are disproportionately affected by shocks and stresses-not only because they are frequently more exposed (and subsequently more vulnerable) to climate-related shocks, but also because they have fewer resources and receive less support to prevent, cope with, and adapt to them (World Bank, 2015).

Experiences from Other Cities

Greater Accra, Ghana: "Integrated Flood Management Makes for a Resilient Metropolis"

With a population of over 4.4 million inhabitants and a growth rate of 3.9%, the Greater Accra Metropolitan Area (GAMA) comprises one of Ghana's most populated areas. Due to high rural-urban migration and the proliferation of informal settlements, the GAMA has expanded to cover 3,245 sq km and 225 km of coastline, encompassing the Accra Metropolitan Area capital district and beyond. The GAMA is prone to coastal and surface inundation, and related water and vector-borne disease epidemics. Rapid urbanisation combined with poor spatial, water and sanitation, and solid waste management compound this vulnerability, resulting in high annual flood risk. In June 2015, heavy rainfall and flooding triggered a catastrophic fire outbreak. Emergency services were unable to respond effectively to the combined flood and fire disaster, that claimed the lives of over 200 people and caused damages amount to USD 55 million (World Bank, 2017). The disaster brought the necessity for *resilience building* and integrated solutions to the forefront.

The reasons for Accra's sustained flood risk are both structural and non-structural. They include poor drainage systems with limited capacity to absorb storm water due to narrow culverts or the collection of solid waste in drains and the prevalence of impermeable surfaces. During flood events, lack of safe and sufficient sanitation facilities increase exposure to water-borne diseases, such as cholera. Outbreaks are reported each year, with the urban poor being particularly affected owing to their location and limited access to health services. Weak spatial planning and control is an additional stressor, as local officials struggle to control/manage the expansion of informal settlements in flood-prone areas or waterways. Population growth is only expected to aggravate the precarious situation (ICLEI, 2017).

Against the backdrop of the deadly 2015 flood, the GAMA collaborated with the World Bank to apply the City Strength Diagnostic tool, that helped assess the cost of the disaster and quickly evolved into a resilience building framework for the metropolis. The City Strength Diagnostic catalysed further collaboration with the private sector, research, and academia and created opportunities for engagement of key community stakeholders in resilience building efforts. The Regional Institute for Population Studies (RIPS) of the University of Ghana with support from the International Development Research Centre (IDRC) is undertaking a three-year research project aimed at addressing socio-demographic change and climate-induced flood risks. This collaboration is expected to strengthen policy-relevant research to support local leaders' decision-making, improve flood management through an integrated smart flood management framework, and enhance inclusive and sustainable urban governance. The project has been designed to ultimately support Ghana's Nationally Determined Contributions (NDCs) and the national Medium-Term Development Planning (RIPS, 2017). In the case of Greater Accra, measures were taken to strengthen planning and governance, enhance the resilience of vulnerable communities by restricting settlement in flood prone areas, improve access to basic services and integrate climate risk management in city planning processes.

City of Cape Town: Building Urban Resilience Through Water

Climate change is a major stress, and, Cape Town is particularly vulnerable to its impacts, that are expected to become more frequent and intense. Cape Town is increasingly characterised by informality, with over 200 informal settlements having been established in the city. The challenges of daily stresses and intermittent shocks are exponentially higher for individuals living in these settlements. Cape Town has applied resilience to its water sector. In 2018, Cape Town made international headlines when faced with the prospect of a potential "Day Zero": a scenario in that the city government would control water supply by turning off certain parts of the reticulation system. The City of Cape Town partnered with engineering consultancy Arup and the Stockholm International Water Institute (SIWI) to develop a global framework for water resilience that will help cities better prepare for, and respond to, challenges to their water systems. Supported by the Resilience Shift and The Rockefeller Foundation, the City Water Resilience Approach (CWRA) uses qualitative and quantitative indicators to measure and build a growing body of knowledge on urban water resilience. By applying the CWRA, cities can diagnose challenges related to water and utilise that information to inform planning and investment decisions. The City identified key action points to building urban water resilience:

- Coherent and relevant governance structures are necessary to promote and embrace urban water resilience. This action aims to create an enabling environment to facilitate the transition from grey infrastructure to majority blue-green water-sensitive design solutions and natural infrastructure, where appropriate.
- Building urban water resilience requires an equitable approach that ensures nobody is left behind. A proposal for the creation of an information platform for each informal settlement in the city, that details water and sanitation issues in each location, that can be used to inform the City's formal budget-setting process.
- Building urban water resilience requires robust partnerships, collaboration, and trust. This action focuses on a comprehensive communication strategy to build trust in government, through enhancing and improving the interaction between government entities, sector stakeholders and citizens.
- Building urban water resilience requires infrastructure that is designed and constructed to enhance resilience in the city. This action proposes a review of the City's existing investment programme to ensure that it is properly resourced, aligned with the new Water Strategy, and considers anticipated water demand and possible long-term growth;
- Building urban water resilience requires that accurate and up-to-date data is available for decision-makers to use. This action seeks to develop a decision support system to enable effective management and optimisation of water resource allocation.

Quito, Ecuador: Transforming into a resilient city by addressing urban mobility.

Quito, the second highest capital city in the world and second largest city in Ecuador is home to over 2.4 million people in its metropolitan area. Climate change is one of the City's most pressing social, environmental, and economic challenges (ICLEI, 2017). Due to its location, the capital is exposed to various geological and hydro-meteorological hazards. Massive seismic movements, floods, and forest fires have tested the City in recent decades. Urban sprawl along steep slopes intensifies exposure to these risks, especially for the poor. Water insecurity is also a growing threat, especially for the agricultural sector.

In 2014, Quito set out to develop an expanded Resilience Strategy with the support of the 100 Resilient Cities programme. The city's context was assessed to understand existing strengths, weaknesses and opportunities to enhance resilience in a holistic way. Quito's transportation system stood out as a promising vehicle for this objective. By targeting the newly expanded public transport system, Quito aims to guide the development of areas built around the metro and Bus Rapid Transit lines in a sustainable, resilient way. The transportation strategy thus becomes part of the city's resilience arsenal. To leverage these benefits, Quito has developed the Eco-efficiency tool, that enables informed decision-making regarding urban sprawl and disaster risk reduction (earthquake and flooding) for new real estate projects.

Lusaka, Zambia

Lusaka has a particularly weak infrastructure endowment compared to other major African cities; ranking 14th out of 20 African cities assessed in a 2014 study by KPMG. This challenge is reflected in the limited coverage of basic hard infrastructure. Wastewater treatment is, generally, accepted as being inadequate with the majority (60%) of water treatment infrastructure beyond its reasonable useful life-time and the few existing water utilities unable to finance full operation and maintenance, let al.one make investments needed to serve a growing population. Solid waste management is also a major problem and, in the context of increasing extreme weather events, poses a significant risk to human health. As a result, Lusaka is suffering from a sanitation crisis that claims lives through annual outbreaks of cholera, typhoid and dysentery (AfDB, 2015).

Based on its vulnerability to climate change and other shocks and hazards, Lusaka has been a ground for urban resilience programming for a long time. For instance, Lusaka City was a beneficiary of the "*Sustainable Urban Resilient Water for Africa: Developing Local Climate Solutions—'SURe Water 4 Africa*", that was implemented from December 2012 to December 2017. The project aimed to contribute to sustainable climate change (CC) resilient urban water planning mechanisms and action based on international benchmarking within local authorities while ensuring multiplier effects to the region. The project focused on the nexus of climate change and water, in particular droughts and floods, while identifying and implementing priority adaptation measures through a participatory planning approach to assist the most vulnerable sectors in the project local authorities. It also developed capacities for resilience planning and implementation at the local level. The UNDP in partnership with UN-Habitat implemented the 'building disaster resilience capacity in Lusaka", that responded to challenges, such as lack of inter-departmental linkages in building to resilience to natural hazards. The programme strengthens the City's capacity to develop and implement resilience plans/strategies.

Major Learning Points from the Cases

Significant lessons emerge from the above cases. First, urban resilience building is not a preserve of local government alone. It requires collaboration from municipalities, private sector, civil society, academia and Non-Governmental Organisations. Second, urban infrastructure (water, sanitation, energy, communications, and transportation systems) is critically important for emergency response and quick recovery of the community and its economy. Where there is vulnerability to a wide range of natural hazards, there are opportunities for enhancing the resilience of critical systems. Third, innovation is key for urban resilience measures. There is a pressing need for new tools and approaches that strengthen local administrations and empower citizens, while building their capacity to face new challenges and better protect human, economic and natural assets. The tools and approaches must be integrated in urban planning and management practices. Third, research and evidence are key for informing decisions on resilience planning.

The State of Urban Resilience in Harare

This section presents and analyses the status of urban resilience in Harare. The focus is on: (i) the vulnerability context, (ii) adoption of resilience focused urban planning and development, (iii) capacities for urban resilience building and (iv) challenges for urban resilience building.

Vulnerability Context

The upper Manyame sub-catchment where the city of Harare lies is already facing erratic rainfall, prolonged droughts and an increase in mean temperatures (Masimba, 2016). Climate projections indicate decreases in precipitation and an increase in temperature in the next forty years (Masimba, 2016). A decrease in rainfall points to the need for coming up with adaptation measures to reduce the impacts of climate change (Masimba, ibid.).

In a survey of 22 sub-Saharan cities, water supply coverage in Harare had declined at a rate of 0.7% from the period of 2001–2012, and the city had 15–30% households that spent 30 min or more fetching water (Hopewell and Graham, 2014). The Service Level Benchmarking (SLB) indicators reveal that service quality continues to decline. This decline in water provision is closely linked to economic decline and illustrates how socio-economic status can affect adaptive capacity of a city (Muller, 2016). It is estimated that Harare requires 1.2 billion litres of water daily, but the city only has the capacity to provide roughly half of that, or 620 million litres. A study conducted by the Harare Residents' Trust in Tafara and other suburbs indicates that taps frequently run dry, and water is often contaminated with sewage and pollutants, forcing residents to use borehole water and unprotected wells (Chatiza et al., 2013). In another study, women in some suburbs of Harare report having not had access to piped drinking water for weeks. The precarious water challenges contribute to disease outbreaks. For instance, from October 2011 to January 2012, there were 1,078 reported cases of typhoid in Harare linked to use of contaminated water from unprotected sources. In January 2017, the Health and Child Care Ministry reported 200 cases of typhoid. The continuous outbreak of water-borne diseases demonstrates failure in urban environmental health and planning (Chirisa et al., 2015).

Urban Resilience Building Efforts in Harare

The vulnerability context described above makes urban resilience a priority in Harare. The question is what is being done to build urban resilience in the city? One of the most recognized programmes is the Harare Slum Upgrading Project (HSUP). HSUP was a five-year collaboration involving the City of Harare (CoH), Dialogue on Shelter for the Homeless People in Zimbabwe Trust (DOS) and the Zimbabwe Homeless People's Federation (ZHPF). The latter two are an Alliance where DOS offers technical support to ZHPF, a social movement of the urban poor affiliated to Slum Dwellers International (SDI). HSUP was supported by the Bill and Melinda Gates Foundation (BMGF). Its implementation was part of BMGF's Global Development Program on Inclusive Municipal Governance (GPIMG) launched in 2010. GPIMG targeted five African cities with an overall goal of building partnerships between city governments and the urban poor to tackle urban development challenges inclusively. The programme prioritised infrastructure improvements in marginalised urban communities (Development Governance Institute (DEGI, 2016). It enhanced access to basic services, such as water, sanitation and hygiene in underserved communities. Urban upgrading prioritises the infrastructure, housing, livelihoods, and social networks of the most vulnerable households living in slum settlements. It promotes socio-economically viable low-income settlements as part of the overall city development (World Bank, 2013). However, the city has not adequately institutionalised slum upgrading in its land use planning policies and practice. Settlement upgrading is still being done using disjointed approaches.

The Government of Zimbabwe, with the support from the United Nations Development Programme (UNDP) and UNICEF is rolling out the "Partnership for Building Urban Resilience in Zimbabwe" (still in pilot phase), whose goal is to improve urban resilience and strengthen the provision of basic social services and Local Economic Development (LED) targeting unemployed youths, women, and vulnerable groups in urban and peri-urban areas.[1] Harare is one of the pilot sites, with particular focus on Budiriro and Glen View low income residential areas. However, there are several factors that hinder effective urban resilience building in Harare.

[1] UNDP. Urban Resilience Building Programme document, 2019.

Factors Militating Against Effective Urban Resilience Building in Harare

Poor Municipal Governance

Resilience is greatly influenced by the quality of urban governance (EU, 2006). Weak and unaccountable city and municipal governments contribute to the lack of basic infrastructure and services, the dynamics of land markets and lack of access to safe land by the poor (World Bank, 2015). Despite the ascendancy of resilient infrastructure in the mitigation of climate change shocks affecting the liveability of the city, the continued political strife undermining efficient service delivery in Harare city for years now seems to have trivialised the urgency of the matter. The deterioration in the delivery of infrastructure and related services in Harare has been attributed to a combination of the poor institutional capacity of the city and its tottering economy (African Development Group, 2011). The local media is frequently awash with news about the rampant cases of corruption in the management of resources ear-marked for service delivery in the deeply polarised city (Dube, 2011). Funding set aside for the building of water treatment works is sometimes diverted to private use by officials in charge of the projects (Mapira, 2011) under the full glare of the serious financial challenges daunting the city. Added to this raft of challenges is the lack of transparency and accountability in the governance of city finances.

Under-Investment in Critical Infrastructure

Under-investment in infrastructure maintenance also contributes to the poor living conditions of urban residents in Harare, characterised by significant infrastructure deficits for basic services: water and sanitation (WASH), waste management, transport, health services, and electricity. The rapid rural–urban migration exerts pressure on urban infrastructure, whose capacity has been exceeded since it was designed for a smaller population (UN-Habitat, 2005). Serious urban infrastructure deficits have huge impacts on poor communities, including slums. Poor infrastructure poses a greater range of threats to cities-and puts obstacles in the way of achieving resilience (World Bank, 2015).

Weak Frameworks for Land-Use Planning and Management

Theoretically, good planning and regulatory frameworks are critical in building urban resilience (Chirisa et al., 2016). However, there are shortfalls in local government and regulatory frameworks in Harare. The institutional framework for municipal adaptation is very weak. Large sections of the urban population and the urban workforce are not served by a comparable web of institutions, infrastructure, services and development control regulations (Dodman, 2009). It is common for between a third and a half of the entire urban population to be living in illegal settlements formed outside any land-use plan. These illegal land uses include squatter or slum settlements and illegal land sub-divisions (Dodman, 2009) mainly in the interstitial spaces of the city and satellite towns of Epworth and Chitungwiza, and the make-shift peri-urban areas of Caledonia and Hopley.

The chaotic land occupations (code named *'jambanja',* vernacular for chaotic violent land invasions) on the fringes of all major towns and cities in Zimbabwe in 2000 triggered the mushrooming of informal settlements-mainly driven by housing cooperatives led by veterans from the country's war of independence from Britain. Since the land occupations were chaotic it is unsurprising that the make-shift settlements developed without prior planning permission, let alone without any form of rudimentary infrastructure in the form of roads, water and sewerage systems as required by the country's planning law. The product is not only the loss of resilient vibrant ecosystems in the greenbelts of most cities in the country but, also frequent outbreaks of epidemics including cholera and typhoid as the desperate new settlers are forced to rely on water drawn from polluted rivers and wells for household uses.

Another dimension is that of outdated Master and Local plans (Berrisford, 2014; Matamanda, 2019). If Master and Local Plans are up to date and authorities developed them following the minimalist participation processes in the current legislation, they would conceivably have better responded to the urban poor's needs than the chaos their absence allowed. Incompleteness of planning processes and products make urban resilience building in Harare tenuous (Muchadenyika & Williams, 2017).

Lack of Climate Change Adaptation Planning

There is a dearth of literature on direct efforts towards climate adaptation in Harare. Rather, there is an indication that there are efforts towards general improved service delivery in the city as opposed to climate adaptation per se. Rehabilitation of water infrastructure has been the main focus of projects in Harare (Mtisi & Prowse, 2012). This is mainly because the city of Harare operates using infrastructure that was established before independence in the 1980s to cater for 300,000 people and yet the population has continued to grow to 1,485,232 people (Nhapi, 2015). Water demand management has also been employed to improve efficiency (Mtisi & Prowse, 2012). However, demand continues to outstrip supply, and this is likely to be worsened by climate change (Mtisi & Prowse, 2012).

There have been a few initial efforts to develop sectoral and local climate adaptation agendas and associated actions within Zimbabwe, including in Harare. One notable example is a consultative process undertaken in November 2017 on the implementation of the National Adaptation Plan (NAP), with the aim of engaging local authorities to improve resilience to climate change (Mutingwende, 2017). The workshop was spearheaded by the United Nations Development Programme (UNDP)

Zimbabwe and the Ministry of Water and Climate. During the workshop, local authorities were urged to embrace climate change adaptation as the impacts of climate change will likely continue to add on to existing problems, such as urbanisation (Mutingwende, 2017).

Conclusion and Recommendations

In conclusion, Harare lacks sufficient funding and revenue sources to develop and implement strategies towards urban resilience. Moreover, the City demonstrates deficient capacity to mobilise, secure, and manage the financial resources needed to meet pressing urban infrastructure investment needs (especially in water and sanitation systems) for climate resilient and sustainable urban development. We hope that the urban resilience programme and the resultant national urban resilience strategy will build a strong case for cities in Zimbabwe to prioritise urban resilience in their planning and development processes. Based on the findings presented in this chapter, the following recommendations are suggested:

- Strengthening urban planning and regulatory frameworks. This requires identifying and enforcing regulatory mechanisms, such as by-laws and planning policies that prevent building and development in hazardous locations, and risk-aware building and infrastructure standards. This action may also include adoption and implementation of integrated urban resilience plans that safeguard vulnerable people's livelihoods and access to basic services. Platforms should also be created for women, youth and other vulnerable groups to inform planning processes, to ensure that planning and implementation are gender-sensitive and respond to their needs and priorities.
- Mainstreaming climate change in city planning efforts. The impacts of climate change are more severe in cities as compared to rural areas. This is mainly because of the concentration of the built environment and people and the reliance of urban dwellers on urban infrastructure systems. As such, climate change should be a cornerstone of the planning and management of cities. In this regard, cities should prioritise the following: (i) Research on vulnerability of cities, (ii) Development of climate change mitigation and adaptation plans, (iii) Design of resilient city infrastructure (energy, water, sanitation, storm water and transport), (iv) design and use of energy-efficient and climate sensitive city energy systems and (v) Internalising national policy statements and strategies on climate change.
- Strengthening the quality of urban governance, promoting community engagement, local-led advocacy to achieve citywide effective policy impact. Increasingly, there is need for urban areas to be able to withstand disasters, shocks and stressors, and progressively build the future resilience of their populations. It is critical to rely on effective and inclusive formal and informal governance structures that operate with accountability and transparency and effectively, at the various levels of the city, to enable cities to be resilient. All governance actors

need to be empowered and capacitated to actively and equitably participate in resilience activities and strategies (from community members to local governments to private actors and service providers). Entry points for this strategic pillar would include (i) providing education, training to and building capacities of local actors, (ii) providing local governments with technical assistance, (iii) influencing regulations and planning principles, (iv) establishing multi-sectoral partnerships, and (v) improving financing.

- Strengthening community-based adaptation. The most important partnership for a resilient city is the one between local government and vulnerable communities. Community-based adaptation may decrease the correlation between poverty and low levels of climate resilience by empowering urban citizens to share and apply their knowledge while influencing inclusive governance practices that integrate these voices and ensure that they are effectively reflected in sustainable outcomes.
- Developing municipal technical capacity. Developing local capacity in risk management and resilience planning must be a key strategy to reduce the multiple risks that Harare and its population are exposed to.

References

African Development Group. (2011). *Infrastructure and growth in Zimbabwe. An action plan for sustained strong economic growth.* African Development Bank Group.

Berrisford, S. (2014). The challenge of urban planning law reform in African cities. In S. Parnell & E. Pieterse (Eds.), *Africa's urban revolution* (pp. 167–183). University of Cape Town Press.

Chirisa, I., Nyamadzawo, L., Bandauko, E., & Mutsindikwa, N. T. (2015). The 2008/2009 cholera outbreak in Harare, Zimbabwe: Case of failure in urban environmental health and planning. *Reviews on Environmental Health, 30*(2), 117–124.

Chirisa, I., Bandauko, E., Mazhindu, E., Kwangwama, N. A., & Chikowore, G. (2016). Building resilient infrastructure in the face of climate change in African cities: Scope potentiality and challenges. *Development Southern Africa, 33*(1), 113–127.

Development Governance Institute. (2016). End evaluation of the Harare slum upgrading project (HSUP) Report. Available online www.degi.co.zw/projects.php

Dodman, D. (2009). *Building urban resilience in the least developed countries.* International Institute for Environment and Development.

Dube, T. (2011). *Systemic corruption in public enterprises in the Harare metropolitan area: A case study.* Unpublished master of public administration thesis. Pretoria: University of South Africa.

ICLEI (2017), Resilient cities report 2017. Tracking local progress on the resilience targets of SDG 11.

Chatiza et al. (2013). Cited in Ndlovu, K., Kagoro, J., & Chatiza, K., (2013). *Report on research on local authorities' capacities to provide services in a gender sensitive manner*. Report Submitted to ActionAid Zimbabwe

Mapira, J. (2011). Urban governance and mismanagement: An environmental crisis in Zimbabwe. *Journal of Sustainable Development in Africa, 13*(6, 2), 258–267.

Masimba, O. (2016). *An assessment of the impacts of climate change on the hydrology of upper manyame sub-catchment, Zimbabwe.* Unpublished M.Sc. Dissertations. Harare: University of Zimbabwe.

Matamanda, A. R. (2019). Battling the informal settlement challenge through sustainable city framework: Experiences and lessons from Harare. *Zimbabwe, Development Southern Africa,*. https://doi.org/10.1080/0376835x.2019.1572495

Meerow, S., Newell, J. P., & Stults, M. (2016). Defining urban resilience: A review. *Landscape and Urban Planning, 147*, 38–49.

Mtisi, S., & Prowse, M. (2012). *Baseline report on climate change and development in Zimbabwe.* Harare: Government of Zimbabwe.

Muchadenyika, D., & Williams, J. (2017). Politics and the practice of planning: the case of Zimbabwean cities. *Cities*, 33–40. https://doi.org/10.1016/j.cities.2016.12.002

Muller, M. (2016). Urban water security in Africa: The face of climate and development challenges. *Development Southern Africa, 33*(1), 67–80.

Murray, M. J., & Myers, A. G. (Eds.). (2006). *Cities in contemporary Africa.* Palgrave Macmillan.

Nhapi, I. (2015). Challenges for water supply and sanitation in developing countries: case studies from Zimbabwe. *Understanding and managing urban water in transition* (pp. 91–119). Springer, Dordrecht.

Pickett, S. T. A., Cadenasso, M. L., & Mcgrath, B. (Eds.). (2013). *Resilience in ecology and urban design, linking theory and practice for sustainable cities.* Springer.

Plan International. (2016). Child-Centred Urban Resilience Framework. *A holistic, systematic and action-based framework for making cities more resilient for children and youth, girls and boys.* Available online https://www.plan.org.au/-/media/plan/documents/reports/curf_brochure2016v8.pdf

RIPS. (2017). *Cities & climate change.* www.rips-ug.edu.gh/index.php/projects/cities-climate-change

UN. (2016). Goal 11: Make cities inclusive, safe, resilient and sustainable. https://www.un.org/sustainabledevelopment/cities/

UN-Habitat (2010). *State of African cities 2010: Governance, inequality and urban land markets.* UN-Habitat.

UN-Habitat. (2016). *Urbanisation and development: Emerging futures. word cities report.* https://unhabitat.org/wp-content/uploads/2014/03/wcr_the-rules-of-the-game-urban-governance-and-legislation-1.pdf

UN-Habitat. (2018). Building sustainable and resilient cities. World Cities Day Concept Note.

UNICEF. (2017). *Strategic note on UNICEF's work for children in urban settings.* Accessed May 07, 2019. http://www.unicefinemergencies.com/downloads/eresource/docs/urban/02.01-strategic%20note%20on%20unicef's%20work%20for%20children%20in%20urban%20settings,%2022%20may%202017.pdf

UNICEF. (2018). Shaping urbanisation for children. *A handbook on child-responsive urban planning.* Accessed May 07, 2019. https://www.unicef.org/publications/files/unicef_shaping_urbanisation_for_children_handbook_2018.pdf

European Union. (2016). Urban resilience.*A concept for co-creating cities of the future. Resilient Europe Baseline Study.* Available online https://urbact.eu/sites/default/files/resilient_europe_baseline_study.pdf

United Nations Human Settlements Programme (UN-Habitat). (2005). Report of the fact-finding mission to Zimbabwe to assess the scope and impact of operation Murambatsvina. UN-HABITAT.

World Bank. (2013). *Building urban resilience: Principles, tools, and practice.* The World Bank

World Bank. (2014a). *The World Bank Annual Report 2014.* Washington, DC. © World Bank. https://openknowledge.worldbank.org/handle/10986/20093. License: CC BY-NC-ND 3.0 IGO.

World Bank. (2015). *Investing in urban resilience. Protecting and promoting development in a changing world.* The World Bank

World Bank. (2017). Enhancing urban resilience in the greater Accra. www.openknowledge.worldbank.org/handle/10986/27516

World Vision International. (2016). *Resilient cities for children: A literature review for world vision's cities for children framework.* Available online: https://www.wvi.org/sites/default/files/Resilient%20Cities%20Summary_Final.pdf

Elmond Bandauko (MPA, BSc. Hons. Rural & Urban Planning) is currently a PhD Student and SSHRC Vanier Scholar in the Department of Geography and Environment, University of Western Ontario, London ON, Canada. He is also a Graduate Fellow with the Center for Urban Policy and Local Governance, one of the affliate Centers of the Network for Social and Economic Trends (NEST) at the University of Western Ontario. Elmond's research interest spans areas such as urban policy and governance, urban transformation in Global South cities (gated communities and new cities), urban informality, housing, urban marginality and urban resilience. His research has been published in *Development Southern Africa, Urban Research and Practice, Urban Forum, Journal of Housing and the Built Environment, International Development Planning Review,* among other journals. Elmond is also a co-editor of the book 'peri-urban developments and processes in Africa, with special reference to Zimbabwe (Springer, 2016).

Chapter 7
Urban Resilience Under Austerity: A Case Study of Street Children Vending in the Harare Central Business District, Post-2018

Witness Chikoko, Samson Mhizha, and Langton Mundau

Abstract The chapter problematises vending as part of urban resilience among the street children of the Harare Central Business District in the face of socio-economic challenges partly induced by the austerity era as from September 2018. The children respond to these challenges by devising various survival strategies. However, very few researches have been conducted to examine the street children's attitudes towards vending as a survival strategy. Research findings suggest that street vending is one of the survival strategies among the street children of the Harare Central Business District, Zimbabwe. The street children demonstrate their agency through selling of sweets, sex, condoms and compact discs, among others. They are able to eke a living in a constraining environment thus demonstrating thin agency. Drawing from a children's rights perspective, street vending by these young people typifies gross violation of their rights in the Harare Central Business District. The qualitative research methodology punctuated by street ethnography was used to generate data for this study.

Introduction

"The recent socio-economic challenges experienced from September 2018 to December 2019" have had a huge toll on human welfare in Zimbabwe. Some of the socio-economic problems have been aggravated by the pronouncement austerity measures by government. The victims of these austerity measures are largely the poor Zimbabweans that include women, men and children. The situation is even worse for children living and working on the streets. However, there seems to be paucity of academic studies that interrogate vending as a survival strategy by children on

W. Chikoko (✉) · L. Mundau
Department of Social Work, University of Zimbabwe, Mt Pleasant, PO Box MP167, Harare, Zimbabwe

S. Mhizha
Applied Psychology, University of Zimbabwe, Mt Pleasant, PO Box MP167, Harare, Zimbabwe

I. Chirisa and A. Chigudu (eds.), *Resilience and Sustainability in Urban Africa*, Advances in 21st Century Human Settlements,
https://doi.org/10.1007/978-981-16-3288-4_7

the streets of the Harare Central Business District within the context of austerity era. The Government of Zimbabwe, through the Finance and Economic Planning Ministry introduced austerity measures as from October 2018 to the present [2019] (Government of Zimbabwe, 2018).

Part of a cocktail of reforms include reduction in Government expenditure, transactions tax, liberalisation of fuel procurement, introduction of interbank foreign currency exchange and reintroduction of the local currency through abolition of the multicurrency regime that was adopted beginning of 2009 during the Government of National Unity (Government of Zimbabwe, 2019). The introduction of austerity measures has been received with mixed feelings. For example, the macro-economic adjustments triggered a runaway inflation that is at 176% from 6% as at September 2018 (*Sunday Mail* dated on 25 August 2019). There has been significant erosion of people's incomes, shortage of basic goods and services, particularly those that require foreign currency for importation. For example, critical shortages of medical drugs, fuel, electricity, bread, among others, has been noted. Ironically, there has been escalation of prices of basic goods and services despite stagnation of people's incomes. In other words, there has been an increase in poverty levels among the citizens of Zimbabwe.

The impact of these austerity measures is even more pronounced among vulnerable children, such as those living in the streets. Therefore, the chapter problematises the vending as a livelihood strategy by the street children of the Harare Central Business District in the austerity era. Dube (1999: 133) defines survival techniques as "the ways by that the children cope with tough realities of street life". Research findings suggest that vending is part of urban resilience where street children of Harare Central Business District engage in to survive in the face of austerity measures. Vending includes various items, such as the selling of airtime, sweets, condoms, aphrodisiac and psycho active substances. Some of the livelihood opportunities include, among others, parking vehicles, begging, selling of several wares and products including sweets, cigarettes, condoms, air-time and aphrodisiac substances, among others. The street children engage in vending in a context that has multiple constraints (Giddens, 1983).

Theoretical Perspectives

This study is informed by two important, though sometimes conflicting conceptual frameworks: child rights and child agency. Bell (2012: 284) defines agency as "a process whereby individuals are able to envisage different paths of action, decide among them and then take action along a chosen route." Chuta (2014: 02) sees agency as "an individual's own capabilities, competences and activities through that they navigate the contexts and positions of their life worlds fulfilling many economic, social and cultural expectations". In simple terms agency is about choices or actions taken by individuals to survive or meet their daily needs (Bourdillon, 2009). Bell (2012: 284) defines sexual agency among youth as a "processes where young people become sexually active and the strategies, actions and negotiations involved in maintaining relationships and navigating broader social expectations." However, some scholars have criticised agency as it sometimes clashes with societal value system as noted by Bordonaro and Payne (2012). For example, Chikoko (2014) observed that commercial sex work among street children of the Harare Central Business District could be regarded as ambiguity of agency. Gigengack (2006, 2008) has also noted that some of the behaviours of street children, such as excessive abuse of substances leads to death, thus showing self-destructive agency. Actions or behaviours, such as gang-rape and substance-abuse among the street children of the Harare Central Business District could be viewed as ambiguous agency. In addition, the actions could also be part of the notions of self-destructive agency because of a number of risks that are associated with the behaviours.

The UNCRC (1989) defines child rights into four principles namely, the best interest of the child, the right of a child to participation, non-discrimination and the right of a child to survival and achieve development. Save the Children (2002) also noted that, child rights perspective recognises the relationship between the duty bearer and the rights holders. Chikoko (2014), Chikoko et al. (2018d) observed that child rights perspective could be seen as social contract that exists between the rights holders and the duty bearers. Nhenga (2008) noted that in an effort to domesticate the provisions of the UNCRC (1989) and the ACRWC (1999), the Government of Zimbabwe has enacted a number of child rights laws, policies and programmes. Chikoko (2014) observed that some of the laws, policies and programmes included the Children's Act (5:06) and Criminal Law (Codification and Reform) Act (9:23) whilst policies or programmes include the Multi-Sectoral Response to Child Sexual Abuse and the National Action Plan for Orphans and Other Vulnerable Children (2016–2020), among others. However, a number of criticisms or flaws have been raised against the UNCRC. Scholars, such as Nhenga (2009), Bourdillon (2009), Morrow and Pells (2012), observed that the UNCRC is seen or viewed as a western conceptualisation of childhood. In addition, Morrow and Pells (2012: 04) noted that "the UNCRC does not contain specific rights relating to poverty and does not define the term." The vulnerability of the children is explained when some of the street children of the Harare Central Business District engage in gang-rape and abuse of substances. In addition, the behaviours are seen as inconsistent with the provisions of the UNCRC, the ACRWC and some of the national child rights laws, policies and programmes. The behaviours demonstrate huge child rights violations prevalent on the streets of the Harare Central Business District.

Research Methodology

Purposive sampling was used to select or identify eight (8) participants for the study. As a result, the researcher identified and targeted participants who were known to be deviant in terms of sexual behaviours and also abusing substances. Through purposive sampling, it was possible to select participants who were considered to be hard to reach. Babbie and Mouton (2012), Neuman (2011) observed that purposive sampling is suitable when working with difficult or vulnerable populations, such as street children and street-based sex workers. The authors added that through purposive sampling, it becomes easier to identify and recruit the participants in a study. Mhizha (2010, 2014) also used purposive sampling on his studies with the street children of Harare. A number of qualitative data collection techniques, such as, life history interviews, in-depth interviews, informal conversations and semi-participant observation methods were adopted to collect data for this study.

Bell (2012) used life history interviews and in-depth interviews when researching about sexual lives of young people in rural Uganda. The qualitative data was analysed through thematic content analysis. Data analysis focused on themes and sub themes that emerged from the study. Some of the themes and sub themes that emerged included, selling aphrodisiac substances, selling psycho active substances and sex, among others. Neuman (2011, 1997) defines ethical considerations as what is or not legitimate to do or what is 'moral' when conducting a research. The following ethical considerations were observed when conducting this research; informed consent, confidentiality and benevolence, among others. In the case of the street children of the Harare Central Business District, the researchers obtained verbal informed consent. The researchers also ensured confidentiality by using alphabetical codes instead of writing the names of the street children on data gathering tools.

Research Findings

The data analysis yielded three themes. The first theme focused on selling of airtime, sweets, condoms, among others. The second theme focused on vending of psycho active substances. Thirdly, the focus is on trading of aphrodisiac substances.

Vending of Pirated Compact Discs (CD), Airtime, Sweets, Condoms

As a result of hardships partly induce by austerity measures, it was evident that some of the adolescent street boys survived through vending of various items. Some of the items included; pirated compact discs, airtime, sweets, vegetables, substances and condoms, among others.

Case One

Tee (not real name) is one of the street boys aged 16. During the informal conversations he indicated that he survives through vending of sweets, condoms, among others. He had this to say:

> Elder, I sell CDs, sweets, cigarettes and condoms. Those are the items that make me survive elder. Elder, in the streets, you do anything, to eke a living.

During the field work visits, the researchers observed that some of the adolescent street boys as social actors were selling cigarettes and condoms. For example, on one of the Saturday nights, the researcher saw Mangongongo (not real name) selling condoms and cigarettes at the entrance of Razzle night club at corner Robert Mugabe and Rezende Street. Some of the Razzle night club patrons also confirmed that they buy condoms and cigarettes from street vendors that include adolescent boys.

Case Two (Selling of Sex, Condoms)

Chipo (not real name) is one of the street girls aged 17 who survives through vending a number of issues, such as sex and condoms on the streets of the Harare Central Business District. Through selling sex and condoms, she is able to take care of the needs of her two children. Her boyfriend who was also a street boy ran away to South Africa to avoid responsibility of taking care of his children.

In addition, Chipo supplements her income though selling sex at various 'bases', such as Njanji and Avenues areas. As a result of selling sex, she would buy food and clothes for her two children. During informal conversations, Chipo revealed the following:

> Elder, life is tough on the streets. None is taking care of our needs. It is ever worse for girls. For me, it is something else. I have two children, who want to eat. For me to survive, I sell a lot of stuff, such as condoms and sweets, among others.

Selling of Substances, Such as Chamba and Chitongo

From the data gathered through field work, it was evident that some of the children of the Harare Central Business District survive through selling of psycho active substances. Some of the psycho active substances included cannabis, blue diamond, *chitongo,* among others. During in depth interviews, one of the adolescent street girls narrated that she sells, cannabis for survival. As a result of her gender, very few law enforcement agents would think that she sells the substances. She added that, because of being a girl, none of the law enforcement agents targeted her for questioning. She had this to say,

> Elder, I survive on selling cannabi or chamba. I get my stuff from jazzmen either from Epworth or DZ. I disguise the law enforcement agents because I am a girl. They never

> think that girls or women can sell chamba. I take advantage of that. I have lots of clients. Some of my clients a commuter omnibus drivers and conductors, fellow street children, sex workers, among others. Sometimes, when I sell the stuff in Zimbabwe dollars (RTGs) I buy foreign currency to store value from inflation. I am able to buy food, clothes and also to have make-up, among other issues.

Selling of Aphrodisiac Substances, Such as Seven Hours, Muchemedza Mbuya etc.

To adapt to their hard realities partly induced by austerity measures, research findings suggest that some of the street children of the Harare Central Business District engage in the selling of aphrodisiac substances. Some of the aphrodisiac substances included, both conventional and traditional ones, for example, "seven hours" and "*muchemedza mbuya,*" among others. During the informal conversations, one of the adolescent street boys had this to say,

> Elder, life is tough these days, I survive through selling aphrodisiac substances. Some of the aphrodisiac substances include, "seven hours". As a result of selling the substances, I am able to buy the basics, such as food and clothing. However, some times, municipal police confiscate some of our stuff. For example, last weekend, all my stuff was confiscated by the municipal police because of the banning of street vending.

During the fieldwork visit, the researchers also observed a number of street children selling aphrodisiac substances. The substances were sold to the general public including to fellow street children. One of the patrons of a popular night club in town confirmed that they buy aphrodisiac substances from street boys. The street boys were selling a wide variety of these substances. Some of these are meant to boost the sex drive of the men seeking them. The patron confirmed that these substances included "wild horse", "seven hours," among others. One of the key informants also confirmed that, the street children were surviving through selling of aphrodisiac substances. She had this to say:

> With some of the challenges confronting the Zimbabwean country, some of the children have resorted to selling of aphrodisiac substances to survive. These children get the tablets, substances from merchandisers as far as Zambia and Congo, among other countries. The children even bargain for fair prices of the substances from these merchandisers.

During key informant interviews, one of the social workers also confirmed that, some of the street children were surviving through selling psycho active substances. Through selling the substances, there were able to raise money for basic needs, such as food, clothes and blankets, among others. However, the social worker also added that, some of the street boys end up abusing the psycho active substances. In addition, the social workers also revealed that other street children end up also using the money they get from selling substances to pay for sexual services with commercial sex workers in town.

Discussion

The above narratives illustrate the agency of the adolescent street children who raise money for survival through vending of condoms, sex, sweets, cigarettes, psycho-active and aphrodisiac substances, among others. The agency is also demonstrated by the ability of these children to negotiate for the pricing of these materials or services. For example, in one of the observations made street children were seen negotiating for fair prices with merchandisers from neighbouring countries, such as Zambia and Democratic Republic of Congo. The ambiguity of agency is also seen as some of the street children survive through selling sex. Bordonaro and Payne (2012) argue that agency becomes ambiguous at two levels. The first level when it threatens the well-being of the society. Selling of sex by street girls is in contrast with the expectations of society. The second level is that when agency affects the well-being of the street children. In the case of street girls selling sex to survive, such agency is ambiguous as the practice exposes the street girls to diseases and vulnerabilities, among other issues.

Some of the actions or behaviours of the street children of the Harare Central Business District constitute thin agency. Tisdall and Punch (2012) define thin agency as practices or social actions that are done within a context characterised with multiple constrains and with limited options. Therefore, given the austerity measures introduced by the Government of Zimbabwe, there were very limited survival options of these children. In such circumstances they ended up with very limited survival options besides engaging in selling aphrodisiac substances, selling of psycho active substances and engaging in transactional sex work, among others.

Some of the practices associated with selling of sex as a survival technique among street girls could be viewed as self-destructive agency. Gigengack (2008) defines self-destructive agency as actions that pose a number of risks to the children. For example, transactional sex work involving street girls could be part of self-destructive agency. Some of the risks associated with transactional sex work among girls include contracting HIV and AIDS and unwanted pregnancy, among others (Chikoko, 2014, 2017).

Drawing from the child rights discourse, the involvement of adolescent street children in vending is viewed as violation of children's rights. The behaviours contravene the provisions of the UNCRC, the ACRWC and some of the national child rights laws, policies and programmes. In other words, vending involving street adolescents is inconsistent with some of the child rights principles. For example, selling of sex by street girls is inconsistent with the 'best interest of the child' principle that is one of the key principles of the UNCRC (1989) and ACRWC (1999). In addition, the involvement of adolescent street boys vending violates the provision of the International Labour Organisations Convention 182 and the Zimbabwe Labour Relations Act (*28.01*). In other words, the involvement of adolescent street children in vending is seen as a 'form of child labour' as defined by the Labour Relations Act (*28.01*).

The engagement of street children in sexual activities, such as transactional sex work is viewed as part of the worst forms of child abuse. It is regarded, as such, because of the circumstances and effects around the practice. In practice transactional sex work involving minors is regarded as sexual abuse, violence and exploitation, among others. For example, there are a lot of risks associated with transactional sex work among street girls. Some of the risks include exposure to sexually transmitted diseases.

Within the child rights concepts, particularly the Criminal Law (Codification and Reform) Act (9.23), street children who engage in selling of psycho active substances, such as cannabis are regarded as 'children in conflict with the law.' In addition, street children who also survive through selling transactional sex are also viewed as 'children in conflict with the law.' This is on the basis of the criminalisation of smoking of cannabis and transactional sex work in Zimbabwe.

In contrast to the *ubuntu/unhu* perspective, the above narrative of selling sex as part of urban resilience among street girls of the Harare Central Business District implies serious moral decadence. It shows that, some of the people of Harare Central Business District; including the street children do not uphold moral values. The realities of transactional sex work and selling of substances among street children are against the values of *ubuntu/unhu.* The behaviour of these children is in contradiction with such values as humanness, love, compassion and kindness, among others as noted by Mangena (2007, 2012), Ramose (2002), Samkange and Samkange (1980). However, when street children do not uphold moral values who should be blamed? It is the society's responsibility that these children are socialised in sound or good moral values. These children were not born by the streets. Scholars, such as Mugumbate and Chereni (2019), Mushunje (2006), observed that it takes the whole village or society to raise a child within the *ubuntu/unhu* thinking/philosophy.

Similarly, Dube (1999) observed that some of the street boys were surviving through vending. He noted that they sell their wares at bus terminus, robots and some were roaming around the city. Dube (1999) observed that some of the items that were sold to the public and the tourists included cigarettes, fruits and confectionery, among others. Mhizha (2010) also noted that street children of Harare raised income for survival through vending activities. He added that those children that were involved in vending were either employed by informal traders or self-employed. The authors also observed that the street children were involved in the vending of a number of items or materials, such as airtime, cigarettes, fruits, vegetables and plastics, among others. However, Mhizha (2010) noted that vending was not confined to adolescent street boys. He cited one of the female street children who were also involved in vending activities.

The study established that as part of urban resilience, the street children of the Harare Central Business District engaged in vending. As highlighted above, they raised money for survival through selling of condoms, sweets, airtime and sex, among others. Similarly, Estrada (2012) observed that, children were very key in terms of supplementing family income through vending related activities.

The research findings are consistent with previous studies by scholars such as Ruparanganda (2008)'s study observed that some of the adolescent street boys devised a number of survival techniques that include selling of parking discs to the general motorists. He noted that the boys were making substantial profits through buying parking booklets from the municipal officials and then selling them to the general public. Similarly, Mhizha (2010) observed that some of the adolescent street girls of Harare were involved in intergenerational sex. He noted that through intergenerational sex, they were able to sustain their lives. Mhizha cited a case of an adolescent street girl who was in love with a white guy and a soldier. "He observed that the sugar daddies" were called "vana Mudyiwa" "Those that are exploited" on the streets of Harare.

The results resonate or concur with studies from Ruparanganda (2008), that observed that some of the street girls survived through intergenerational sex. He cited a case of a street girl who was in love with a certain sugar daddy. Ruparanganda (2008) noted that as a result of the relationship, the girl left the streets and was staying in one of the high-density areas of Warren Park. He also observed the street girl had a number of valuables as a result of, such as a state-of-the-art mobile phone and intergenerational sex.

Ruparanganda (2008) observed that when business people were relocating, they normally engage the adolescent street boys to assist in offering their labour. He noted that the boys would load and off load vehicles with heavy materials that included stationery, furniture and clothing materials, among others. Ruparanganda (2008) also observed that supermarkets and food outlets also sought the services of the adolescent street boys. Andre and Godin (2014) noted that children were very crucial in supplementing family income through artisanal mining related activities. The authors also observed that during the mining process in Democratic Republic of Congo, there was division of labour based on gender and age. Andre and Godin noted that the older boys were involved in activities such as digging while the girls were for crushing rocks, sifting and also cleaning the bags.

The study established that some of the street children of the Harare Central Business District raise money for survival through selling of psycho active substances such as *chamba, chitongo,* among others. Through the selling of the psycho active substances, the street children were able to raise money for survival thus, meeting their day to day basic needs. However, there are risks that, some of the children end up misusing the substances. Other children also interact with criminals that also buy the psycho active substances. There is high probability that some of the children's behaviour could be negatively influenced by the criminals that they interact with when selling substances. The research findings also suggest that some of the street children of the Harare Central Business District eke a living through selling of aphrodisiac substances. As highlighted above, the aphrodisiac substances were wide and varied. However, there are also risks associated with the practices of selling aphrodisiac substances. For example, some of the children are enticed to abuse of the substances. In such circumstances, there are a lot of risks such as nose bleeding as a result of using aphrodisiac substances. Ruparanganda (2008) questioned issues around dosage associated with the use of "*guchu*" concoction as one of the traditional substances used by street children of the Harare Central Business District.

Conclusion and Recommendations

The chapter concludes by arguing that vending is one of the survival strategies of the street children of the Harare Central Business District in the face of socio-economic challenges. Some of the challenges may partly be blamed on the recent move by the Government of Zimbabwe to introduce austerity measures. The austerity measures were introduced as from September 2018 up to present. As discussed above, some of the vending included various items or materials such as sex, condoms, sweets, airtime, aphrodisiac and psycho active substances. The vending of various items demonstrates agency of the street children of the Harare Central Business District. On the other hand, vending by the street children also highlight the level of vulnerability of these children. Drawing, from a child rights perspective, the vending as a survival demonstrates huge child rights violations that is prevalent on streets of the Harare Central Business District. Some of the vending practices are inconsistent with international, regional and national child rights laws, policies and programmes. A number of recommendations have been raised and these include the following:

- There is need to empower street children of the Harare Central Business District. The empowerment can take different dimensions. For example, one of the dimensions would involve access to information. Part of the information can empower street children on proper condom use to reduce risks associated with unprotected sex. Another dimension of empowerment could include the use of social protection such as harmonised cash transfer to fight childhood poverty related issues. The empowerment can also take the dimension of social and vocational skills. The street children can sell some of the skills to earn money for survival,
- There is need for a full implementation of child rights laws, policies and programmes to reduce risks associated with street vending. The full implementation of the child rights laws, policies and programmes will significantly reduce risks associated with vending among street children,
- There is need for the duty bearers to provide decent survival or livelihood opportunities for street children, for example, having Drop-in centres on the streets. These Drop-in centres should provide food hand outs and laundry services to the street children.

References

African Union. (1999). *African Charter on the Rights and Welfare of Children,* Unpublished Addis Ababa.

Bell, S. A. (2012). Young people and sexual agency in rural Uganda. *Culture, Health & Sexuality: An International Journal for Research, Intervention & Care, 14*(3), 283–296.

Cheney, K. E. (2012). Killing them softly? Using children's rights to empower Africa's orphans and vulnerable children. *International Social Work, 56*(1), 92–102.

Chikoko, W. (2014). Commercial 'sex work' and substance-abuse among adolescent street children of Harare Central Business District. *Journal of Social Development in Africa, 29*(2), 57–80.

Chikoko, W. (2017). *Substance-abuse among Street Children of Harare: A Case of Harare Central Business District,* Unpublished D. Phil Thesis, Department of Social Work. Harare: University of Zimbabwe.

Chikoko, W., Chikoko, E., Muzvidziwa, V. N., & Ruparanganda, W. (2016). Nongovernmental organisations' response to substance-abuse and sexual behaviours of adolescent street children of the Harare Central Business District. *African Journal of Social Work, 6*(2), 58–64.

Chuta, N. (2014). Children's agency in responding to shocks and adverse events in Ethiopia. *Young Lives: An International Study of Childhood Poverty* Working Paper 128.

Dube, L. (1997). AIDS-Risk patterns and knowledge of the disease among street children in Harare, Zimbabwe. *Journal of Social Development in Africa, 12*(2), 61–73.

Dube, L. (1999). *Street children: A part of organised society* [Unpublished D.Phil. Thesis Department of Sociology]. Harare: University of Zimbabwe.

Gigengack, R. (2008). How street children studies can address self-destructive agency. In R. Gigengack (Ed.), *Focaal, special issue on contemporary street ethnography,* (Vol. 36, pp. 7–14).

Gigengack, R. (2006). *Young, Damned and Banda: The World of Young Street People in Mexican City 1990–1997* [Unpublished D.Phil. Thesis]. Netherlands: University of Amsterdam.

Mangena, F. (2012). *On ubuntu and redistributive punishment in Korekore-Nyombwe culture: Emerging ethical perspectives*. Best Practices Books.

Mangena, F. (2007). *Natural law ethics, Hunhuism and the concept of redistributive justice among the Korekore-Nyombwe people of Northern Zimbabwe: An ethical investigation* [Unpublished D.Phil. Thesis, Faculty of Arts]. Harare: University of Zimbabwe.

Mhizha, S. (2010). The self-image of adolescent street children in Harare [Unpublished M.Phil. Thesis]. Harare: University of Zimbabwe.

Mhizha, S. (2014). Religious self-beliefs and coping vending adolescent in Harare. *Journal of Religion and Health, 53*, 1487–1487.

Morrow, V., & Pells, K. (2012). Integrating children's human rights and child poverty debates: Examples from in ethiopia and India. *Sociology, 46*(5):906–920.

Mugumbate, J., & Chereni, A. (2019). Using African *Ubuntu* theory in social work with children in Zimbabwe. *African Journal of Social Work, 9*(1), 27–34.

Neuman, W. L. (2011). *Social research methods: Qualitative and quantitative approaches* (7th ed.). New York: Pearson Education Inc.

Nhenga, T. C. (2008). Application of the international prohibition on child labour in an African context: Lesotho, Zimbabwe and South Africa [Unpublished D.Phil. Thesis, Department of Public Law]. Cape Town: University of Cape Town.

Ramose, M. B. (2002). African philosophy through Ubuntu. Harare, Mond Books Publishers.

Ruparanganda, W. (2008). *The sexual behaviour patterns of street youth of Harare Zimbabwe in the era of the HIV and AIDS pandemic* [Unpublished D.Phil. Thesis]. Harare: University of Zimbabwe.

Samkange, S., & Samkange, T. M. (1980). *Hunhuism or Ubuntusim: A Zimbabwe indigenous political philosophy*. Graham Publishing.

Tisdall, K., & Punch, S. (2012). Not so new? Looking critically at childhood studies. *Children's Geographies, 10*(3), 249–264.

United Nations. (1989). *The United Nations convention on the rights of a child*. Unpublished. United Nations.

Chapter 8
Resilience-Based Interventions to Street Childhood Among Street Children in Zimbabwe

Samson Mhizha, Blessing Marandure, and Witness Chikoko

Abstract The chapter explores the efficacy of resilience-based intervention strategies that are aimed at tackling street childhood in Zimbabwe. Resilience-based interventions are those interventions that make use of the inherent capacity and agency of the concerned children. While street childhood is a significant challenge in Zimbabwe and Africa, there are little debates on and use of resilience-based interventions for street children. Zimbabwe is devoid of innovative and well-researched interventions to tackle street childhood. In recent years, resilience has been deployed to explain both the aetiology and interventions to street childhood. Indeed, resilience is explained to influence the decision of getting to the streets by the street children as they attempt to handle family adversity at home. In the same vein, interventions that are based on resilience are suggested in the current study. Analysis of secondary data was conducted for the study. The results reveal that the resilience-based interventions may include religious networks, peer support groups, schooling, use of expressive arts, family and parenting clubs for their parents and guardians, contextual psychosocial support and counselling, vocational training, drug rehabilitation and psychosocial support and counselling.

Introduction

In this chapter, the authors discuss the possibility of use of intervention strategies on street children in Zimbabwe. This goes well with the fledgling phenomenon of urban

S. Mhizha (✉)
Department of Applied Pyschology, University of Zimbabwe, P.O Box MP 167 Mount Pleasant, Harare, Zimbabwe

B. Marandure
School of Applied Social Science, Faculty of Health and Life Science, De Montfort University, Leicester, UK

W. Chikoko
Department of Social Work, University of Zimbabwe, P.O Box MP 167, Mount Pleasant, Harare, Zimbabwe

I. Chirisa and A. Chigudu (eds.), *Resilience and Sustainability in Urban Africa*, Advances in 21st Century Human Settlements,
https://doi.org/10.1007/978-981-16-3288-4_8

resilience. Urban resilience refers to the capacity of urban institutions, individuals, systems and communities to recover and maintain their function and thrive in the aftermath of a stress or shock, regardless of its impact, magnitude or frequency (Frantzeskaki, 2016). This concept of urban resilience, that is interdisciplinary and has gained traction over the recent decades, is very relatable because we are living in an urban age where urban communities, individuals and systems are facing both greater opportunities and challenges that are unprecedented in human history. To cope with the challenges of urbanisation, involving the influx of one million people to the cities, the world would need to build one city weekly, otherwise many people will continue to live in conditions with serious vulnerabilities (Aravena, 2016). Indeed, the concept of urban resilience arises from the observation that the population of those living in urban areas has increased exponentially, leaving many children vulnerable. These children suffer from multiple deprivations; their families do not have access to land, housing or basic services.

Some scholars have reasoned that street children do reveal resilience (Malindi & Cekiso, 2014; Malindi & Machenjedze, 2012; Theron & Malindi, 2010; Theron & Malindi, 2011). However, other scholars have also highlighted that these street children are also suicidal (Kidd, 2004; Kidd & Kral, 2002; Oppong Asante, 2016; WHO, 2010), and that they engage in unprotected and precocious sex in addition to engaging in various other high-risk sexual behaviours for survival (Sorre & Oino, 2013; Oppong Asante, 2016). Research findings also indicate that street children engage in high rate, precocious and pernicious substance-abuse (Oppong Asante, 2016; Pluck, 2015), and are physically abused and stigmatised due to their street childhood (Oppong Asante & Meyer-Weitz, 2015). All these phenomena increase the risk of HIV infection among street children, especially in the context of a high HIV epidemic (Oppong Asante, 2016). Furthermore, street children have also been reported to have poor mental health, partly due to their challenging lifestyles (Aptekar & Stoecklin, 2014; Pluck, 2015).

Interventions to Promote Resilience

There have been various interventions that seek to address the needs of street children in low-income countries. The significance of addressing mental health problems in children who have been exposed to trauma and violence in many poor countries, such as Zimbabwe, is increasingly recognised (Benjet, 2010). The negative outcomes of traumas and stressors in childhood are well studied. Survivors of the childhood traumas and stressors are at an elevated risk for improper sexual behaviours, anxiety, guilt, anger, shame, depression, post-traumatic stress disorder, and other behavioural and emotional problems throughout their lives (Finkelhor et al., 2005; Nanni et al., 2012). Interventions to mitigate such negative outcomes inadvertently address different aspects of resilience in these children. The following is a brief review of such interventions, ending with recommendations for implementation of resilience-based approaches for street children.

Cognitive Behavioural Therapy (CBT)

Cognitive Behavioural Therapy is a potential avenue for dealing with challenges faced by street children in Zimbabwe. CBT is employed to assist children to grapple with self-destructive beliefs and replace them with adaptive and helpful beliefs (Geldard & Geldard, 2008). Use of CBT has the most empirical support of all for mental health interventions (Beck, 2005; Kazak et al., 2010). Interventions based on CBT principles commendably address a broad range of children's emotional and behavioural needs (Butler et al., 2006; Kazak et al., 2010). Specifically, CBT effectively works in trauma-focused therapy with children (Cohen et al., 2006, 2012), both in group and individual situations (Jaycox, 2004). CBT is based on the principle that emotions, thoughts and behaviours are all interconnected. Therefore, when an intervention can change a person's emotions and thoughts, behaviour will also change in line with those newly formed emotions and thoughts. Rather than merely changing behaviours, CBT seeks to change thoughts that underlie behaviours. Therefore, the individual's behaviour being changed by a therapist guarantees that individual will be empowered to change their own emotions, thoughts and behaviours (Butler et al., 2006). Though it has much empirical support, not much research has been done to determine the efficacy of CBT on children, especially those living in the streets in Zimbabwe. It would be important to adapt CBT with cultural, religious, and societal settings of children in Zimbabwe during its implementation on street children in Zimbabwe. Studies have shown that CBT-based interventions are effective in developing children's adaptive coping and proactive problem-solving skills; teaching relaxation skills; reducing anxiety/depression; and strengthening parent-child relationships (Cohen et al., 2006; Jaycox, 2004). In a Zambian study, Murray et al. (2013) showed that through the use of cognitive behavioural therapy activities such as home-based visits, psychosocial counselling and other social activities, therapists were able to handle the multiple negative impacts of stress and trauma in children.

CBT is the only psychological intervention that has been recommended by the World Health Organisation (2004) for use in children and adolescents. According to Kudenga (2016), only less than half of street children he interviewed believed that the mental health interventions they received were appropriate. It is our contention that an adapted version of CBITS (Jaycox, 2004) would be particularly beneficial for street children in Zimbabwe, owing to the limited resources in terms of availability of mental health professionals. The intervention could thus be administered by trained lay persons, increasing feasibility of application in resource limited contexts.

Bibliotherapy

Bibliotherapy was developed by Heath and Heath (2008) on the principle that carefully selected good stories have the power to create changes in attitudes leading to changes in behaviour. Thus, individuals exposed to good stories are motivated to change their thinking, leading to change in their feelings ultimately leading to change in their behaviours (Heath & Heath, 2008). Thus, wisely selected stories build children's adaptive coping and prosocial behaviours (Heath & Cole, 2012; Tukhareli, 2011; Wood et al., 2012). Bibliotherapy is based on Forgan's (2002) conception that articulates that reading stories helps people to deal with emotional or psychological wounds and scars. Teachers and parents can read books to children so that they can build adaptive coping skills (Forgan, 2002). Benefits of bibliotherapy include reducing children's feelings of seclusion, decreasing stress and tension; providing chances for problem-solving strategies; and empowering children to pursue newly acquired insights (Forgan, 2002).

As the children engage in bibliotherapy, they get interested in it and start to identify with major character in the stories that helps the children to see parallels with their experiences (Forgan, 2002). The children start to position themselves in the character's experiences and roles emphasising their thoughts and feelings. At the end of the story, the children internalize the primary message leading to shifts in their thinking and attitudes and they start applying the newly acquired insights into their daily lives, leading to changes in behaviour. Ensuing discussions on the story offer children with chances to further apply their newly acquired insights, leading to reinforcing of the connection between the story's central message and its personal relevance. Donald (2013) offers stories that have been found relevant and helpful for orphaned children in South Africa, basing on his long work in South Africa's educational system.

Bibliotherapy has been identified as efficacious in addressing various mental health issues such as anxiety in children (Rapee et al., 2006). However, in attempting to expose street kids to print based stories, it is imperative to take into account low rates of literacy in such a sample, who miss out on schooling whilst living on the street. As such, application of bibliotherapy would require significant adult input by way of reading the stories to the street children, or providing them with audio-books. Nevertheless, this presents its own challenges, such as the potentially prohibitive cost of audio book supply.

Read-Me-to-Resilience

Stories such as parables and fables, have been utilised over the years to pass on proven teachings from one generation to another (Friedberg & Wilt, 2010). Also, Wood et al., (2012) publicized their preliminary findings based on a piloted Read-me-to-Resilience intervention based on children's stories. Their study was conducted with 32 children and showed that using traditional resilience-themed, African stories were valuable in bolstering the Orphans and Vulnerable Children (OVC)'s resilience. The findings showed that the intervention strengthened resilience and positive outcomes among the children. Furthermore, participating teachers who read these traditional stories to their pupils also showed that they strengthened the resilience of the pupils. These resilience stories were prepared in many languages including English, isiXhosa, and isiZulu.

Culturally Aligned Rites of Passage

Dream work is a potential avenue for dealing with challenges faced by street children in Zimbabwe. Thamuku and Daniel (2012) revealed implementing an intervention that involved culturally aligned rites of passage on orphans in Botswana. The orphaned children had publicly expressed traumatic experiences of loss and grief. The children had also proactively identified and confirmed strengths upon that they could draw personal support that involved strong communal sense of strengthening resilience. Thus, adults must determine the importance of all children feeling included and having opportunities to share life experiences. Additionally, Clacherty and Donald (2008) revealed that the efficacy of the Siyabakhumbula programme that provided a clear process and structure to aid therapeutic grieving elated to HIV/AIDS. This intervention was based on culturally significant practices such as group sharing, ritualistic fire offerings and songs to the deceased relatives. In their findings, they further showed that the intervention aided the children in grieving and accessing their memories and feelings in situating there was otherwise no space and time, to advance rituals of recalling that contained the unbearable and sustained hope. This helped to unearth the collective and inner resources necessary in turning the struggle for survival into an affirmation of life.

Suitcase Project

Another programme implemented in South Africa by Clacherty's (2008) is the Suitcase Project that outlines a supportive process that provides displaced children opportunities to show their memories and feelings whilst telling their stories. Based on this intervention strategy, she authored a book showing 11 children's individual stories (Clacherty, 2008). The book has stories of the displaced children, whose traumatic backgrounds and trajectories led them to cities. Clacherty shows the supportive relevance of urging children to deal with their traumatic grief by developing symbolic artwork involving drawing and painting pictures on the inside and outside of small suitcases that represented their life experiences. Clacherty (2008) showed that when children decorated their suitcases, they simultaneously and slowly revealed their distressing life stories. In the process and over several weeks and months, the Suitcase Project's artwork helped in focusing on each child's story that allowed details to unfold slowly. Gradually, children added transitional objects and mementos and journal entries to the contents of their suitcases. Some children chose different paintings and drawings on the outside of their suitcase that were public and available to all than they elected for the inside of their suitcase that were invisible to outsiders and private. The figurative nature of these drawings fostered ways to gradually unravel their stories at a comfortable pace. Developing the artwork together with sharing personal stories that were linked with the artwork allowed the creation of secure emotional distance from real-life experience. This promoted the sharing of otherwise unreachable and overwhelming traumatic life experiences. Thus, the benefits of such a child appropriate and centred approach are clear to see. Additionally, this would be a low-cost initiative that would address the emotional needs of street children through allowing self-expression.

Memory Box

When working with orphaned children, Ebersöhn et al. (2010) outlined the need to equip community volunteers with intervention strategies that offer supportive focus and value to children's resilience. Ebersöhn et al. (2010) offered narrative therapy that helped children to slowly share their stories through the development of the making of a memory box. This intervention strategy involves encouraging the children to share their stories or narratives that also involved drawing upon positive memories about the deceased loved ones. This approach is based on tenets of positive psychology, that supports children and adults to focus on resources, assets and capacities of survivors (Eloff et al., 2007; Seligman & Csikszentmihalyi, 2000). Since the memory box is decorated, the children describe their memories and tell stories related to the objects they choose to include in it. Special mementos, pictures, official documents, poems, special letters, small trinkets, and other sentimental and prized objects are placed in the box. In a class set up, teachers could urge their pupils to decorate a box. Each box would be exceptional to its creator and may be associated with many emotional attachments, depending on the individual child's losses and experiences. The children's memory boxes may then contain meaningful drawings, keepsakes, notes, poems, among others. For these children, the chosen objects are associated with memories of their deceased loved ones that assist them in showing grief and/or remembering memories of better times. Teachers would then urge the pupils to tell their stories about the selected contents. Over time, and as trust is developed, the children would describe their life stories, increasingly sharing more details to develop their own individual narratives. However, this intervention strategy could potentially be limited by the loss of personal effects during the children's time living in the streets. Nevertheless, children can be encouraged to create substitutes (e.g. through drawings and crafts) for lost items they deem significant for telling their life stories.

Dream Work

Dream work is a potential avenue for dealing with challenges faced by street children in Zimbabwe. Historically, some cultures, like Native Americans in the United States, the Chinese, and Latin Americans would regard dreams as an avenue for meeting their deceased family members and relatives. Such dreams involving the deceased may be an avenue to have insights and solutions for challenging problems, insights into how to handle difficulties, clarity on life's meaning, and a sense of direction and purpose (Bulkeley, 1997). Cluver and Gardner (2006) averred that orphaned children are more likely than their peers to have constant nightmares. Research shows that children's dreams are impacted by their daily experiences (Cooper, 1999). For instance, Ablon and Mack (1980) noted that some of the most common developmental stressors for children can be clearly represented in the content of their dreams—such as changes in parent's emotional states, deaths, illnesses, injuries, frightening events of the day, and similar traumas and stressors. Likewise, Cooper (1999) revealed that grieving children remember their dreams more often and vividly than non-grieving children.

Dreamwork may assist grieving children interpret events and experiences in their lives. Healy-Romanello (1993) showed that grieving children given opportunities to share their dreams and nightmares by talking about them, drawing pictures representing those dreams, or by modelling and depicting their dreams with clay helped them identify and express their painful feelings. Thus, dreamwork may assist children in increased problem-solving, identifying developmental and emotional challenges and solving them.

Creative Expression and Writing

Creative expression and writing may also help children to deal with emotional challenges. Poetry, writing journal entries, creative writing, and engaging in narrative drama or theatre may help children identify and resolve emotional and mental challenges (Nyawasha & Chipunza, 2012). It is important to allow the children to choose their preferred activities. For instance, some children may prefer drama over poetry in dealing with their painful experiences. Thupayagale-Tshweneagae et al. (2010) showed that some children avoided using direct terms related to death and AIDS with one child referring to her mother's AIDS-related death as 'heavy', yet she did not mention the word AIDS. It has been shown that creative expression and writing helped the children deal with challenging experiences.

Play is one avenue through that resilience-based mental health interventions can be offered. UNICEF (1998) suggested that through use of dance, theatre, creative writing and oral storytelling, among other interventions, negative effects of traumas and stressors may be reduced. Play is a strategy through that emotional wounds can be represented through their symbolic importance (Francoise, 2005). Play can be used to identify, bandage and heal the children's emotional wounds. Play allow the children to identify the intellectual and emotional aspects of everyday life (Francoise, 2005). In other words, play allows the children to identify, interpret, experience and handle their emotional challenges under an expert who can help them deal with them.

Parenting as a Catalyst for Change

Positive parenting has been suggested as an effective intervention on street children. Indeed, the evidence that supports parenting interventions in improving the wellbeing and health of children and parents is both expanding and compelling. Parenting interventions have proven effective in reducing violence against children, reducing the risk of anti-social and violent behaviour among children, reducing family stress, and improving maternal mental health (Knerr et al., 2013). Furthermore, positive parenting during early childhood is associated with mitigating the effects of poverty on children's mental and behavioural development and is predictive of lesser involvement in substance use behaviours, and a greater life expectancy (Knerr et al., 2013). As both of these are widespread within the street community, this offers yet another impetus for programming that focuses on strengthening parenting skills.

Beyond the resounding evidence in favour of positive parenting, front line workers and the street youth themselves have identified parenting skills as a skill set that has both the potential to improve their current circumstances, and to break the cycle of street involvement. Indeed, parenting interventions have tremendous potential, as the original programme has demonstrated a reduction in child abuse, caregiver depression and stress, and substance use by caregivers and adolescents (Cluver & Gardner, 2007). It has also led to an increase in parental supervision, household economic status, and social support (Cluver & Gardner, 2007). However, addressing parenting in street children is complex, and involves reuniting them with their parents. For example, children may have run away due to abuse, and thus reunions risk re-traumatising those in such situations. Moreover, others may have been orphaned; hence, there are no parents to reunite them with. Finally, some may have drifted so far from home at a young age that tracking their parents may prove to be an impossible task.

Music

Children's interest and involvement in music suggests it is an important and ongoing part of their environment. By engaging within music throughout their lives, children gain recognition for their strengths (for instance, playing instruments, singing, and rapping) and develop important coping mechanisms that provide them with alternatives to acting impulsively and hanging out in unsafe environments. This suggests that children's early and ongoing relationship with music provides them with an important form of ecological resilience. Fraser et al. (2004) note that environment plays a significant role in resilience, arguing that environmental contexts may facilitate higher functioning through exposure to opportunities for growth, thereby engaging existing protective and promotive factors and developing new ones. In their framework, the symbiotic relationship between environmental assets (e.g. children's early exposure to music) and individual attributes (e.g., children's ongoing interest and involvement in music) produces an ecological resiliency, where existing and newly developed protective and promotive factors may ameliorate risks (e.g., children using their interest and involvement in music to develop alternative coping mechanisms to resist impulsively acting on anger).

Additional findings suggest that children experience the music studio as a space to engage in music production, education, and appreciation. Music production provides children with important opportunities to engage their music-based strengths while working through relational challenges that provide them with important opportunities to develop intra- and interpersonal skills. Music education and appreciation also provide children with opportunities to engage their strengths. Every individual has strengths (Saleebey, 2013). Yet, the majority of youth homelessness literature frames young people experiencing homelessness from a risks and consequences perspective. While it is important to identify youth populations at increased risk of experiencing homelessness and engage in research and prevention efforts to ameliorate the consequences they experience while homeless, it is also important to honour and recognise their strengths. Doing so challenges the dominant risks and consequences narrative in youth homelessness research. It contributes to the growing research on homeless youths' strengths, and to what Kidd (2012) refers to as an empowerment perspective, whereby young people experiencing homelessness are understood as navigating and surviving within societal oppression and discrimination by demonstrating resilience and strength.

Peer Groups

Peer Outreach-Street Work Project is significant in that it serves a population that is often difficult to reach. Homeless and out-of school youths are often missed by traditional interventions. These youths, who are out of the mainstream, may not have access to basic health services and information shared in schools, through the media, or by community-based organisations. The homeless youths, who participate in more in-depth educational interventions and skill development activities with their peers, are able to exert positive pressure on members of their "community" to participate in safe sexual and drug use activities. Ultimately, this will reduce the number of new HIV infections among these street children. Many interventions on street children do not have outreach components to serve hard-to-reach populations. The Teen Peer Outreach-Street Work Project is particularly unique in that it matches teen peer educators with adult outreach staff members to provide street outreach services to homeless youths. Every effort is made to use true peers and not merely youngsters who are the same age as the target population. Many of the youths participating in the project are former clients. They are familiar, first hand, with the issues and needs of the target population.

Many street outreach projects serving the homeless population do not provide in-depth interventions; they merely distribute HIV prevention materials. While this is an important aspect of HIV prevention outreach, it may not affect the target populations' belief structure that is more likely to influence changes in HIV-related risk behaviours. The Teen Peer Outreach-Street Work Project is a norm-changing programme, providing skill development and support. The project identifies natural leaders in communities of homeless youth who then assist in providing on-going education and skill development. The programmes participating in the Teen Peer Outreach-Street Work Project operate shelters for homeless youth. These shelters become a place where HIV prevention messages, delivered on the streets, can be reinforced. Finally, in the Teen Peer Outreach-Street Work Project, diversity is valued and nurtured, with clients and staff members from numerous ethnic, cultural, and sexual communities welcomed. The project provides forums for client and peer involvement in programme design, implementation, and evaluation. The fact that young people are capable and should be given the chance to assume certain staff functions is recognised. The Teen Peer Outreach-Street Work Project provides youth with the knowledge, skills, and encouragement necessary to make productive, healthy choices for themselves.

Spiritual Guidance

Street children must be free to attend churches of their own, participate in prayer meetings and receive spiritual guidance and counselling. Some of the children would have experienced traumatic experiences during their parents' illness period and subsequent death, that some may need spiritual counselling. Ngwenya (2015) reports that spiritual counselling is effective in helping children in difficult circumstances in grappling with the problems they may face, some of which they may not fully understand. Ngwenya (2015) wrote that spiritual counselling has a positive impact because some of these children are too young to understand why they are in the situations they are in, looking after themselves without an adult. Counselling sessions help the children deal with their anxiety, fears and challenges. The NGOs in the study by Ngwenya (2015) based their interventions on Biblical foundations and guided by biblical scriptures and principles from Galatians 5 verse 13 and Jeremiah 29 verse 11 respectively. The findings of this research highlighted the innovative strategies by both Family Aids Caring Trust (FACT) and Simukai in providing psychosocial support.

Ngwenya (2015) also observed involvement of churches from the inception of the NGOs that helped children in provision of spiritual guidance and counselling. Churches have contributed towards sustainability of OVC intervention by providing their church facilities to be used free for meetings by OVCs and their care givers. The churches also partnered with the NGOs in providing community networks for donations of cash or in kind for OVCs within their churches and beyond. The findings of this study are also in line with Omwa and Titeca (2011) who asserts that community safety nets can be churches whose mandate is to cushion OVCs and other vulnerable groups from the shocks of the death of parents and severe economic challenges.

Family Counselling

When OVCs have released their emotions during art therapy and counselling sessions, and they are ready for reunification with their family's key informant Tapiwa had this to share: "Before a child could be reunited with his or her family, Simukai engages family about expressed interest by their child of returning home. The family is counselled on the child's current behaviours, identified challenges of the child and how the child has undergone counselling himself or herself. The family's acceptance of the child and what it means living with the child who has lost family values while on the streets starts the process of rehabilitating OVC from the street and reintegration of the child into the family and school." Sharing her personal experience on benefits of family counselling Caregiver 5 shared the following: "My children ran away from home when my family shack was destroyed by the government during the 2004 "Murambatsvina" [refuse trash] era when we were left homeless. I never knew where my children had gone until Simukai staff identified me through my children in the streets. I was told that my children were at Simukai Safe House and wanted to come back home. I was counselled together with one of my daughters who had remained with me on how to start a new life with my children who had been on the streets for four years and how to help them get back to school. Today I am a caregiver with 134 Simukai because of my children and the counselling I received from Simukai."

Livelihoods Support

A critical avenue of supporting and providing resilience to street children involves providing livelihoods support. Ngwenya (2015) reported that some NGOs cushioned OVCs and their guardians from economic shocks faced by Zimbabwe through some sustainable livelihood interventions. These also involved income savings, lendings, stationery, food hampers and school uniforms. The beneficiaries were trained to be self-sustainable and self-reliant. The families and children may also need training on book keeping, get livestock support and be supported through community projects such as community nutritional gardens and Zunde RaMambo. It would also be important to offer the street children life skills training in areas such as welding, carpentry, sewing and hairdressing. Such skills training may also need project support through provision of capital and exposure to markets. The livelihoods support may also involve getting funding from donors, having a pool of volunteers with long time service and relationship and involving the wider community in OVC projects planning. These OVCs choose what they think they can embark on in business and are trained in that field and on business management principles. After training, OVCs are then given some start-up loan to do their business. For example, some are into chicken feeding, chicken egg production, motor mechanics or hairdressing. Others are also into drama/plays and music. All these activities help OVCs in their sustenance as they continue to have income and can further help some of their family members without the help of NGOs. Being able to sustain their family livelihoods also has the advantage of reducing the likelihood of them returning to the streets.

Conclusion and Policy Recommendations

In the study, the authors interrogated the interventions that can be implemented to help street children in Zimbabwe. Though the interventions are debatable, it is argued that the government departments and academic institutions should collaborate in developing innovative solutions to help the same children. In the study, it has been argued that locally-relevant and contextual interventions are necessary in helping these children. Such interventions are beneficial not only for the children themselves, but for wider society, as they reduce the likelihood of these children returning to the streets, and thus the associated societal problems this creates. The authors recommend that the government increases efforts to ease socio-economic challenges in the country to, especially, help the children who seem to bear the brunt of the crises.

- There should be a framework to help families and guardians to better provide emotional and economic care for their children.
- Street children must also get education. Such education should include training in life skills and vocational training that should impart competences, like decision-making, interpersonal skills, stress and coping with it, health literacy and emotional intelligence.
- Regarding vocational skills, the individual street children and former street children should be free to select courses they want from an array including mechanics, welding, farming, sewing, embroidery, brickwork, carpentry, computer literacy and programming among many others. Government should avail funding for such education.
- Perhaps more importantly, government should provide rehabilitation and therapeutic services to street children to help them deal with the adverse psychological and emotional experiences both at home and in the streets.
- There should be advocacy to deal with stigma, labelling and stereotyping that the street children face in schools and communities.
- Parents, families and communities should also be trained in provision of care and positive parenting for children that should help them prevent, mitigate and detect the cases of street childhood. The training should make the parents aware of the negative and traumatising effects of the behaviours that throw the children into the streets.
- Finally, it is also critical that the government, non-governmental organisations and academics come together to develop and pilot test new and innovative strategies to develop resilience and sustenance in the street children in Zimbabwe.

References

Ablon, S., & Mack, J. (1980). Children's dreams reconsidered. *Psychoanalytic Study of the Child, 35,* 179–217.

Aptekar, L., & Stoecklin, D. (2014). Children in street situations: Street children and homeless youth. In L. Aptekar, & D. Stoecklin (Eds.), *Street children and homeless youth: A cross-cultural perspective* (pp. 5–61). New York: Springer.

Aravena, A. (2016). Two billion more people will live in cities by 2035. This could be good – or very bad. The Guardian Retreived 20 August 2019 on https://www.theguardian.com/cities/2016/oct/19/two-billion-more-people-live-cities-alejandro-aravena-habitat-3.

Beck, A. T. (2005). The current state of cognitive therapy: A 40-year perspective. *Archives of General Psychiatry, 62*(9), 953–959.

Benjet, C. (2010). Childhood adversities of populations living in low-income countries: Prevalence, characteristics, and mental health consequences. *Current Opinion in Psychiatry, 23*(4), 356–362.

Bulkeley, K. (1997). *An introduction to the psychology of dreaming*. Westport, CT: Praeger.

Butler, A. C., Chapman, J. E., Forman, E. M., & Beck, A. T. (2006). The empirical status of cognitive-behavioural therapy: A review of meta-analyses. *Clinical Psychology Review, 26,* 17–31.

Clacherty, G. (2008). *The suitcase stories: Refugee children reclaim their identities.* Cape Town: Double Storey Books.

Clacherty, G., & Donald, D. (2008). *Evaluation of the Ekupholeni Siyabakhumbula project, 2007–2008.* Johannesburg: Ekupholeni Mental Health Centre.

Cluver, L. D., & Gardner, F. (2006). The psychological well-being of children orphaned by AIDS in Cape Town, South Africa. *Annals of General Psychiatry, 5*(8), 1–9.

Cluver, L., & Gardner, F. (2007). Risk and protective factors for psychological well-being of children orphaned by AIDS in Cape Town: A qualitative study of children and caregivers' perspectives. *AIDS Care, 19*(3), 318–325.

Cohen, J. A., Mannarino, A. P., & Deblinger, E. (2006). *Treating trauma and traumatic grief in children and adolescents.* New York, NY: Guilford Press.

Cooper, C. A. (1999). Children's dreams during the grief process. *Professional School Counselling, 3*(2), 137–140.

Donald, D. (2013). Positives and caveats of a potentially seminal book: A review on the handbook of African educational theories and practices; a generative teacher education curriculum by Nsamenang and Tchombe (2011). *Journal of Educational Research: Reviews and Essays, 1*(1), 26–30.

Ebersöhn, L., Eloff, I., & Swanepoel-Opper, A. (2010). 'Memory boxes' as tool for community-based volunteers. *Education as Change, 14*(1), S73–S84. Retrieved From http://137.215.9.22/bitstream/handle/2263/16156/ebersohn_memory%282010%29.pdf?

Eloff, I., Ebersöhn, L., & Viljoen, J. (2007). Reconceptualising vulnerable children by acknowledging their assets. *African Journal of AIDS Research, 6*(1), 79–86. https://doi.org/10.2989/16085900709490401.

Friedberg, R. D., & Wilt, L. H. (2010). Metaphors and stories in cognitive behavioural therapy with children. pp. 9—173.

Finkelhor, D., Ormrod, R., Turner, H., & Hamby, S. L. (2005). The victimization of children and youth: A comprehensive, national survey. *Child Maltreatment, 10*(1), 5–25.

Forgan, J. W. (2002). Using bibliotherapy to teach problem solving. *Intervention in School and Clinic, 38,* 75–82.

Francoise, D. (2005). *The combat for the child welfare gives a direction to our humanity.* Paper presented at the conference of Children and Wars, Palace of Luxembourg, Luxembourg.

Frantzeskaki, N. (2016). *Urban resilience: A concept for co-creating cities of the future. Resilient Europe* [Concept Paper]. Erasmus University.

Geldard, K., & Geldard, D. (2008). *Counselling children: A practical introduction.* London: Sage.

Healy-Romanello, M. A. (1993). The invisible griever: Support groups for bereaved chil-dren. *Special Services in the Schools , 8*, 67–89.

Heath, C., & Heath, D. (2008). *Made to stick: Why some ideas survive and others die.* New York, NY: Random House.

Heath, M. A., & Cole, B. V. (2012). Strengthening classroom and teachers' emotional support for children following a family member's death. *School Psychology International, 33*(3), 243–262.

Jaycox, L. (2004). *Cognitive behavioural intervention for trauma in schools (CBITS).* Longmont, CO: Sopris West.

Kazak, A. E., Hoagwood, K., Weisz, J. R., Hood, K., Kratochwill, T. R., Vargas, L. A., et al. (2010). A meta-systems approach to evidence-based practice for children and adolescents. *American Psychologist, 65*(2), 85–97.

Kidd, S. (2012). Invited commentary: Seeking a coherent strategy in our response to homeless and street-involved youth: A historical review and suggested future directions. *Journal of Youth and Adolescence, 41*, 533–543.

Kidd, S. A. (2004). The walls were closing in, and we were trapped–A qualitative analysis of street youth suicide. *Youth and Society, 36*, 30–55.

Kidd, S. A., & Kral, M. J. (2002). Suicide and prostitution among street youth: A qualitative analysis. *Journal of Adolescence, 37*, 411–430.

Knerr, W., Gardner, F., & Cluver, L. (2013). Improving positive parenting skills and reducing harsh and abusive parenting in lowand middle-income countries: a systematic review. *Prevention Science, 14*(4), 352–363.

Kudenga, M. (2016). Appropriateness of counselling given to street children by non-governmental organisations in Harare, Zimbabwe. *International Journal of Scientific and Research Publications, 6*(1), 507–513.

Murray, L. K., Familiar, I., Skavenski, S., Jere, E., Cohen, J., Imasiku, M., et al. (2013). An evaluation of trauma focused cognitive behavioural therapy for children in Zambia. *Child Abuse and Neglect, 37*(12), 1175–1185.

Nanni, V., Uher, R., & Danese, A. (2012). Childhood maltreatment predicts unfavourable course of the illness and treatment outcome and depression: a meta-analysis. *American Journal of Psychiatry, 163*(2), 141–151.

Ngwenya, M. (2015). *An investigation into challenges faced by community-based interventions for orphans and vulnerable children in Mutare, Zimbabwe. Master of Arts.* University of South Africa.

Nyawasha, T. S., & Chipunza, C. (2012). An assessment of psychosocial and empowerment support interventions for orphans and vulnerable children in Zimbabwe. *Journal of Human Ecology, 40*(1), 9–16.

Omwa, S. S., & Titeca, K. (2011). *Community-based Initiatives in response to the OVC crisis in North Central Uganda.* Belgium Institute of Development Policy and Management: Uganda.

Oppong Asante K., & Meyer-Weitz, A. (2015). Association between perceived resilience and health risk behaviours in homeless youth. *Journal of Adolescence, 39*, 36–39.

Oppong Asante, K. (2016). Street children and adolescents in Ghana: A qualitative study of trajectory and behavioural experiences of homelessness. *Global Social Welfare, 3*(1), 33–43.

Pluck, G. (2015). The 'street children' of Latin America. Retrieved on 12 July 2018 from https://thepsychologist.bps.org.uk/volume-28/january-2015/street-children-latin-america.

Rapee, R. M., Abbott, M. J., & Lyneham, H. J. (2006). Bibliotherapy for children with anxiety disorders using written materials for parents: A randomized controlled trial. *Journal of Consulting and Clinical Psychology, 74*(3), 436.

Saleebey, D. (2013). Introduction: Power in the people. In D. Saleebey, (Ed.), *The strengths perspective in social work* (6th ed., pp. 1–23). Upper Saddle River: Pearson.

Seligman, M. E. P., & Csikszentmihalyi, M. (2000). Positive psychology: An introduction. *American Psychologist, 55*(1), 5–14.

Sorre, B., & Oino, P. (2013). Family based factors leading to street children. Phenomenon in Kenya. IJRS, India, 2(3), 148–155.

Thamuku, M., & Daniel, M. (2012). The use of rites of passage in strengthening the psychosocial well-being of orphaned children in Botswana. *African Journal of AIDS Research, 11*(3), 215–224.

Thupayagale-Tshweneagae, G., Wright, S. D., & Hoffmann, W. A. (2010). Mental health challenges of the lived experiences of adolescents orphaned by HIV and AIDS in South Africa. *Journal of AIDS and HIV Research, 2*(1), 8–16.

Tukhareli, N. (2011). Bibliotherapy in a library setting: Reaching out to vulnerable youth. Partnership. *The Canadian Journal of Library and Information Practice and Research, 6*(1), 1–18.

UNICEF. (1998). Cultural media to attack the problem of the children touched by the war, the models possible. Symposium conducted at the meeting of the *Ministry for Foreign Affairs and the international Trade of Canada*. Vancouver, Canada.

Wood, L., Theron, L. C., & Mayaba, N. (2012). 'Read me to resilience': Exploring the use of cultural stories to boost the positive adjustment of children orphaned By AIDS. *African Journal of AIDS Research, 11*(3), 225–239. https://doi.org/10.2989/16085906.2012.734982.

World Health Organisation. (2004). *Prevention of mental disorders: Effective interventions and policy options: Summary report*. Geneva: WHO.

World Health Organisation. (2010). *Antiretroviral Drugs for treating pregnant women and preventing HIV infection in infants: Recommendations for a public health approach 2010 version*. Geneva, Switzerland: Author. Retrieved From http://whqlibdoc.who.int/publications/2010/978 9241599818_eng.pdf.

Chapter 9
Childcare and Familial Relations During the Migration Period

Thebeth R. Masunda and Pranitha Maharaj

Abstract Migration of parents, especially the mother, has greater effects on a family's living arrangements in the country of origin. Due to migration, care arrangement for children is reconfigured where an alternative caregiver is identified for the migration period. This chapter examines the effects of childcare on relationships between migrant parents and non-migrant caregivers. A qualitative approach was adopted where 60 interviews were conducted between 30 migrants and their 30 non-migrant family members. Findings of the study revealed that although migrants identify care-givers they had good relations with before migration, misunderstandings on parenting approaches during the migration period affected their relationships negatively. There seems to be a thin line separating discipline and abuse, that in the end causes tension. What one may refer to as discipline and training children to be responsible, someone else may view as abuse. To avoid conflicts, some caregivers turn a blind eye to the mischievous behaviour of the children, that then causes children of migrant parents to be regarded as ill-mannered. It is therefore, crucial for migrants and non-migrant caregivers to find a common ground on parenting approaches so that the children can get the care and nurturing they require during the migration period.

Introduction

Migration presents both opportunities and challenges for societies and individuals as many people, especially from low-income countries, adopt it as a means of earning a living. The increased movement of individuals in the contemporary migration age

T. R. Masunda (✉)
Department of Community and Social Development, University of Zimbabwe, Mt Pleasant, PO Box MP167, Harare, Zimbabwe
e-mail: ru6masunda@gmail.com

P. Maharaj
School of Built Environment and Development Studies, University of KwaZulu-Natal, Durban, South Africa

I. Chirisa and A. Chigudu (eds.), *Resilience and Sustainability in Urban Africa*, Advances in 21st Century Human Settlements,
https://doi.org/10.1007/978-981-16-3288-4_9

has greatly impacted on the trends and patterns of migration and its effects. The International Organisation for Migration (IOM) (2017) highlights that such changes in the trends and patterns of migration have changed the face of migration in general. Although migration is commended for creating livelihood opportunities for individuals and their families, there are social problems that emanate from the same migration process (Mukwembi & Maharaj, 2018; Safta et al., 2014). In the contemporary migration, individuals of varying age groups and from different backgrounds adopt migration as a livelihood strategy. Konseiga (2006) argues that migration is a household decision to secure better livelihoods and improve family or household well-being.

Migration can therefore be regarded as a family decision in the context of risk aversion. As such, both positive and negative effects of migration are felt by the whole family or household. Research has indicated that both migrants and their non-migrant family members are affected by migration, although the impact varies depending on a number of factors (Smith et al., 2004; Schmalzbauer, 2004; Gibson et al., 2011). Worth noting is that individuals from diverse social background and of different age groups engage in migration (Mukwembi & Maharaj, 2018; Safta et al., 2014). As such, individuals who are still in their reproductive ages are also embarking on migration. Some manage to migrate with their children and spouses while others leave them in their countries of origin. In the event that children are left back home, migrant parents make childcare arrangements to ensure that their offspring have someone to look after them during this migration period. Research shows that children left behind by migrant parents benefit in terms of increased consumption, better access to health and education facilities from the remittances transferred (Ratha et al., 2011; Safta et al., 2014; Smit, 2001). However, the same scholars noted that separation of children, especially minors from their parents can have some negative effects on the child's welfare.

Selection of caregivers during the migration period should be carefully done so that the rightful person is chosen for ensuring better welfare and upkeep of the children left behind. Using qualitative data from a PhD study on exploring the effects of migration on familial relations between migrants and their non-migrant family members, this chapter explores issues surrounding child care during the migration period. The chapter is tailored to evaluate the effect of childcare on relationships between migrant parents and the family members they entrust their children during the migration period.

Migration and Childcare During the Migration Period: A Review

The movement of people from one country to another has increased over the years. Demurger (2015) posits that millions of people worldwide live and work outside their countries of birth. Given the economic disparities, migration has become an essential response to structural disequilibria between and within sectors of an economy (Adepoju, 2006). In this era of deepening globalisation, migration is impacting all countries and individuals in some way (Schmalzbauer, 2004; IOM, 2017). Trends and patterns of migration are rapidly changing and, in turn, this changes the world and migration in general (IOM, 2017). Movement of labour between countries is an important component of growth and development with varying effects both in the sending and receiving countries (IOM, 2005; IOM, 2017). Although migration creates the opportunity to improve the livelihood for individuals and families thereby enhancing their social and economic well-being, it also has the tendency to accentuate problems associated with family disintegration (Ratha et al., 2011; Safta et al., 2014).

Gibson et al. (2011) state that the impact of migration on the migrants and individuals left behind appears to be multifaceted. On one hand, migration enhances welfare of the left behind family members by improving their consumption and risk-copying ability through remittance transfer (Gibson et al., 2011; Crush et al., 2018; Crush & Tevera, 2010). On the other, since migrants are part of a family unit, their departure is likely to create gaps in the day-to-day functioning of the families and households that they were active in prior to migration. The duties and responsibilities that they had before migration are either shifted to the remaining family members or left unattended to (Hoang et al., 2015, Gibson et al., 2011). Such changes cause disintegration and disruption within the family and household structure. The migration of a family member, specifically parents, is a process that transforms the family relationships and its functioning (Graham & Jordan, 2011). The degree of adjustment required afterwards depends on a variety of factors. Such can depend on who is the migrant family member, age and gender of those left behind and the period and frequency of migration (Hoang et al., 2015, Gibson et al., 2011).

Increased labour movement, especially the feminisation of such movements suggests that more children are growing up in transnational families (Hoang et al., 2015). The search for work takes many migrants away from home. As such, transnational family has become the main feature of the contemporary migration matrix. Children are geographically separated from their parents over an extended period of time. A large number of children are growing up without either or both of their parents due to migration. Such has necessitated the kinship support systems for raising the children and providing care. Baldassar et al. (2007: 13) states that:

> Geographic distances between family members provide empirical evidence stressing out that mobility is a common feature among contemporary kinship groups and that members of families retain their sense of collectivity and kinship in spite of being spread across multiple nations.

As parents migrate, they leave a member of the family to care for their children, thereby strengthening their social ties. Mukwembi and Maharaj (2018) argue that migration has the potential to strengthen ties between family members through different transnational activities.

Although family structures and functioning are disrupted due to parental migration, Hoang et al. (2015) and Smith et al. (2004) argue that such disruptions are more pronounced when the mother is also a migrant. In the event that fathers migrate, mothers continue with their socially ascribed roles of caring and nurturing their children while maintaining the existing nuclear family structure (Hoang et al., 2015). They also take over the tasks previously performed by men, thereby adding more duties and responsibilities (Demurger, 2015). As such, little or no change in the living and care arrangements is observed if the mother stays in the country of origin. However, when the mother migrates, many remnant fathers seek help from the extended family members, friends or neighbours to undertake caring and nurturing tasks (Hoang et al., 2015). This also applies in the scenario where both parents migrate; the children left behind are likely to move in with the extended family with either the children moving to the extended family household or an adult from the extended family household moving into the migrants' household to look after the children.

In the event that both parents migrate, care arrangement for children is reconfigured (Graham & Jordan, 2011). Childcare provided by someone else other than the parent is now commonplace. Parents who work away from home depend on alternative care networks to assist raise their children during the migration period (Schmalzbauer, 2004). The caring and nurturing role of biological mothers is shifted to grandmothers, sisters, daughters, aunts, neighbours and friends. These kin members are crucial pillars for the family survival during the migration period. Smith et al. (2004) argues that it is through such networks that mothers can migrate, because without someone to look after their children, women would not be able to migrate.

The feminisation of migration has prompted anxieties about care crisis and the future of the family in the sending countries. Smit (2001) indicates that as much as migration offers access to a livelihood in a different geographical location, there are possible social problems that may be encountered by both migrants and their non-migrant family members. The absence of one or both parents during the child's socialisation can have negative effects on his/her well-being and upbringing. Literature highlights that migration can cause social vulnerability to members left behind (Smit, 2001; Dissanayake et al., 2014). Moreover, the migration of one spouse may also strain the marriage. Temporary circular migration also increases the risk for family breakdown and fragmentation of social networks (Demurger, 2015).

The longer the periods of separation between parents and children, the more children lose parents' reference, their authority and roles as providers of love and care (Dissanayake et al., 2014). Parents are gradually replaced by other family members or children take upon themselves the task of parenting. Such separation has long-term consequences in the children's lives. Migration of the parent is associated with academic, health, behavioural and emotional problems for children left behind (Antman, 2012; Mu & de Brauw, 2013). In rural China, Mu and de Brauw (2013) found that child care became scarcer in migrant-sending households. Due to increased responsibilities, less time is prioritised for cooking and monitoring the eating habits and behaviours of children. As such, many children of migrant parents had stunted growth. However, these effects are said to vary depending on a number of factors.

Although migrant parents make childcare arrangements before they migrate, existing literature provides a fairly negative view of substitute care arrangements for children left behind by migrant parents (Hoang et al., 2015; Smit, 2001). Such care arrangements are viewed as inadequate since children of migrant parents show feelings of loss and abandonment. Children have emotional, belonging, and intimate relationship issues (Graham & Jordan, 2011). Caregivers have a critical role to play in the development and socialisation of children left in their care. Dissanayake et al. (2014) highlight that the development and socialisation of any child depends on the quality of care they receive. It depends on the adults who notice when they are hungry or sick and are able to meet those needs. They argue that good care entails proper interaction between the child and the care giver. Drawing from such arguments, it is critical that the rightful individuals are selected as caregivers of children left behind by migrant parents.

Study Area

The study was carried out in two countries; Durban, South Africa and Harare, Zimbabwe. Initial study interviews were conducted with Zimbabwean migrants living in Durban and the second phase of interviews was carried out in Harare with the respective family members of the interviewed migrants in Durban. Durban is the largest city in South Africa's KwaZulu-Natal Province and is one of the country's main seaside resort cities. It is located at the far-east side of the country, about 600 km from Johannesburg. Harare is the capital city of Zimbabwe, situated in the north-east of the country, in Mashonaland Province. As the capital city, it is Zimbabwe's leading financial, commercial and communication centre with residents from diverse cultures and backgrounds.

Research Design

The research entails a qualitative study of Zimbabwean migrant workers in Durban and their respective family members back home in Zimbabwe. Creswell (2009: 4) defines qualitative research as "a means for exploring and understanding the meanings individuals or groups ascribe to a social or human problem". Qualitative research uses concepts and clarifications to interpret human behaviour and social phenomena from the perspective of the people affected by the phenomena under review (Cohen et al., 2002). Hennink et al. (2010) argue that the main distinctive feature of qualitative research is that it allows the researcher to identify and understand issues from the perspective of the participants. The emphasis is on the verbal description of the phenomena in its natural setting. This study sought to explore relationships during the migration period; thus, a qualitative research design was the most suitable method as it allowed for the interpretations and meanings that people attach to migration to be explored.

The study employed a snowball or chain sampling technique. Hennink et al. (2010) state that snowball sampling is a recruitment method suitable for identifying study participants with specific characteristics or information required in a study. This sampling technique relies on personal contacts to recruit study participants. For the purposes of this research, a total of 60 interviews were contacted, where 30 interviews were done in Durban and the other 30 in Harare. The researcher identified two migrants from Harare who then identified others who had family members in Harare as well. The process was repeated until a total of 30 migrants were interviewed in Durban, where 16 were males and 14 females. Out of the 16 male migrants interviewed in Durban, 75% were married, 20% were single and 5% were not married but cohabiting. One married male participant was in a polygamous relationship with three wives.

Of the total number of female participants, 35.7% were married, 35.7% were widowed, 7.1% were cohabiting and 21.4% were single. 81.2% of the male participants had children while 68.8% of female migrants reported that they had children. Most of the participants interviewed in Harare were family members of the migrant participants in Durban. Of the participants in Harare, 40% were parents of the migrants while 33.3% were siblings. Extended family members constituted 20 and 6.7% were hired caregivers. Data was collected using two types of interviews in qualitative data-eliciting methods: narrative accounts and one-on-one in-depth interviews. These two types of data collection tools are important when the researcher wants to elicit detailed views from the participants in a short period of time. Participants gave narrative accounts of their lives and experiences of migration that were unguided by research questions. This narration was followed by one-on-one in-depth interviews in that research questions were used to elicit the specific pieces of information required to answer the identified research questions.

Narrative interviews entail the retelling of stories from the informant's point of view (Jovchelovitch & Bauer, 2000). Hence, the stories that are told and the way they are told reflect the understanding of the person telling it, that might be different from the way the next person may relate the same story. This makes the narrative inquiry a suitable methodology since it allows for the participant's actual interpretation of the issue at hand to be revealed. After engaging in narrative discussions, the researchers conducted in-depth interviews with the participants. Miller and Glassner (1997) highlight that one-on-one in-depth interviews, in a qualitative enquiry, provide the researchers with the opportunity to explore the participants' point of view regarding the issue at hand. As such, people's beliefs, perceptions, feelings and emotions, and the meanings they attach to experiences will be learned in the process. Given that this study sought individual meanings and perceptions, in-depth interviews were a relevant means of collecting data. Allowing the participants to narrate their stories first then engage in interviews later allowed the researchers to elicit more information from the participants. Issues were raised in the narratives that the researchers followed up on during the in-depth interviews for further clarity, hence, obtaining the relevant information required to answer the research questions. The research design and data-eliciting methods adopted in this study ensured that participants' voices were heard. Hence, conclusions of the study are based on the information provided by the study participants.

Findings of the Study

In evaluating and understanding the effects of social, economic, political and cultural changes in Zimbabwe, the researchers found that many households lament the continuous movement of their families and homes as a result of migration. Both migrants and their non-migrant participant family members believed that had the political, economic and social outlook of Zimbabwe been stable, they would not have opted for migration. However, given the economic disparities between and among countries and regions, families spread their human capital over geographically dispersed and structurally different markets to reduce risk and maximise the chances of securing their livelihood. This study, like many others conducted on migration effects (Demurger, 2015; Hoang et al., 2015; Smith et al., 2004), revealed that, as members of different families and households, migrants have particular duties and responsibilities in the day to day running of the households. Their departure as a result of migration can have notable effects on the well-being of their families. Moreover, the study revealed that not all migrants are able to migrate with their children and other dependents. Most of the migrant participants reported that they had one or more children staying in Zimbabwe. Financial capacity and need for children to be socialised using familiar cultural values were reported as major reasons for leaving children behind. Such information is analysed against the background of child care and its effects on relationships during the migration period.

Caregiver Selection During the Migration Period

Family has been identified as one of the main sources of social and economic support, especially in rural Zimbabwe. Social support is, generally, defined as the assistance that is provided to individuals through social relationships and interpersonal transactions, while economic support is the assistance in monetary value (Barr, 2004). As individuals go through their daily experiences, they are stressed by a variety of life events, hence the need for both social and economic support. Participants in the study revealed that they tend to draw both social and economic support from their family members. Female migrants, especially those who left children in Zimbabwe, reported that their migration journey was successful because they managed to make arrangements for a family member to look after their children during the migration period. Such results are in support of Hoang et al. (2015) who argue that children of migrant mothers are likely to move in with an extended family household or an adult extended family member moves in the migrant's household to look after the children. One female participant who is a divorcee reported that,

> It was my sister who took my daughter into her home, otherwise I would not have managed to come to South Africa. My daughter did not have a passport and also it was not going to be possible for me to bring her even if she had the passport because I did not have a job, I was staying with some friends.

Another widowed female participant reported that she only managed to go to South Africa after her mother in-law took her two children in. Such sentiments support Smith et al. (2004) who argue that migration of mothers can only take place once careful childcare arrangements are made in the country of origin. Unlike the migration of a male figure in the household or migration of the father, a mother's migration is met with a lot of challenges due to her caring and nurturing responsibilities. For a mother to migrate, there is need for a rightful person to assume her caring and nurturing roles in the family to ensure that the welfare of the children is not compromised. It was, however, highlighted that this separation of children and parents has greatly affected household living arrangements. It has also affected the relationships between migrants and the family members, either those left in the care of the children or those who wanted to be left with the children but were not given that privilege. Both migrants and their non-migrant family members reported that childcare during the period of migration has strengthened some relationships, while others were weakened, and some were completely broken down.

Migrants reported that choice of the caregivers was not random; there were certain criteria that they used to identify the rightful people to look after their children in their absence. Findings from this study indicated that relationships before migration played a bigger role in the selection of caregivers during the migration period. Migrants reported that they left their children in the care of someone they had good relations with before migration. One of the female migrant participants, regarded her sister, as the only person suitable to look after her child in her absence.

> Even if I am to die today, she will automatically adopt my daughter. She is the only person who has had my back from the time I was a child, and I could not think of anyone else to entrust with my child's life.

She said that there was no way she would have left her child with her husband's family given the treatment she received from them before migration. Given her relationship with her sister, she knew that she was the right person to look after her child. Most of the female participants in the study shared similar sentiments. They believed it was risking their children's welfare to leave them with someone who disregarded them (mothers) before migration. They argued there was no guarantee that such a person will have their children's interests at heart. Most of them commended:

> How can one care for a child when they hate that child's mother? It's mostly likely the hatred will be transferred to the child.

There were some instances where circumstances compel migrants to find an alternative person, other than the one they prefer, to leave their children with during the migration period. A married female participant who stays with her husband in South Africa reported that although she wanted her children to stay with her mother, it was not favourable for her to do that given that her mother stayed in a rural area and schools were far from home.

> Although I wanted my children to stay with my mother, the problem is that schools are very far from her home and it was going to be difficult, especially for my four-year-old who is in ECD, to walk twelve kilometres to and from school every day. I used to walk these distances when I was young, but I couldn't let my children go through that. As a compromise, I had to hire a childminder to stay with them in Harare. In the end, they remain in their own home.

She also reported that having a hired childminder to look after her children in her own home, rather than to take them to her mother's rural home was a blessing in disguise because it shielded her from gossip. She highlighted that given her relationship with her in-laws, sending children to her mother was likely to create a wave of rumours.

These opinions were shared by other participants who indicated that sometimes, the selection of the person to look after children during the migration period was marred by gossip and accusations of favouritism, given that when someone is looking after the children, the migrant parents will always remit to the caregiver, and whatever is remitted will not only benefit the child but the caregiver as well. Some family members would want to look after migrants' children so that they can get such benefits. One participant reported that:

> Since my sister is looking after the children, my mother-in-law is telling everyone that it was a plan that I came up with so that I would send money to my sister and not to her. Yet sometimes my sister uses her own money to cover for some of the children's requirements. When I tell my mother-in-law that sometimes we go for two months without sending money to the children, she does not believe it.

However, some migrants admitted that their choice of caregivers is influenced by the fact that they want the children to be with someone that they have a strong bond with, someone they can send their money to freely. One female migrant openly admitted that, at times, one has to be strategic when making such decisions.

> We have to admit, whoever we leave the children with will also benefit from our sweat and tears, so I would not give that benefit to someone who is not very close to me.

The participant admitted that she chose her mother to look after her children over her husband's sister because she did not want her sister-in-law to benefit from the remittances that she would send to the children. She reported that the sister-in-law used to treat her badly when she was still in Zimbabwe:

> They never wanted me to marry their brother from day one. They used to eat all food without giving me a thing, so how can I let them benefit from my sweat now? After all, what guarantee do I have that they will not eat the food and not give my child as they used to do for me?

Such reports show that not every family member can be a caregiver during the migration period. Migrant parents, especially mothers, relied on the strength of social relations to identify caregivers (Mukwembi & Maharaj, 2018).

Relationships and Childcare During the Migration Period

Regardless of the relations and ties that existed before migration, that most of the migrant participants attribute to the selection of the caregiver, the study revealed that there are some parenting discrepancies between migrants and their non-migrant family members. The different parenting techniques between migrants and the caregivers were reported to have caused a rift between them. The issue of disciplining children was reported as a major cause of tension between migrant parents and family members selected as caregivers during the migration period. Participants reported that if not well handled, child discipline could ruin relationships that were in existence before migration. It was reported that there seems to be a thin line separating discipline and abuse, that in the end causes tension. What one may refer to as discipline and training children to be responsible, someone else may view as abuse.

A 40-year-old migrant widow, reported having problems with entrusting her children to her family members, and that affected their relationships. She had this to say,

> I first came to South Africa in 2006, soon after my husband's passing. I worked for about a year and a half, but my children were not ok, so I decided to go back home. I had left them in the care of my brother and his wife. I received reports that my sister-in-law was ill-treating my children and I decided to go back and look for someone else to look after them. When I left, she seemed happy to look after the children, but things did not work out. Every time I visited, I would ask the children if everything was okay and they would tell me shocking stories, so I decided to go back home. It's really tough to entrust your children to someone else.

She felt that her sister-in-law was turning her children into housemaids, as she made them do everything even though there was a hired maid at home. She was surprised that her brother supported his wife, as they both said it was good to train the children to be responsible, rather than spoil them. Although she admitted that it was good to train children to do house chores, she did not, like overdoing it because it would then become abuse.

> Yes, there is nothing wrong with teaching children to do housework, but it's another thing to wake them up every day at 4 am so that they can clean the house before they go to school but there is a housemaid employed to do that.

The woman reported that she has now left the children under her young sister's care. However, the sister who is now looking after the children reported that her elder sister is overprotective of her children and that makes it difficult to discipline them because they will cry abuse. She said that given the interaction she has had with her elder sister now that she is staying with the children, she no longer believes earlier accusations that their sister-in-law was abusing or ill-treating the children. Rather, she believes it is her sister who has a problem with another person disciplining her children. She reported how they always argued over disciplining the children. Every time she reprimands the children, they would complain to their mother, and that caused them to fight more often. The caregiver reported that at one point she tried to discipline her sister's daughter after she came home very late but she told her mother and there was a confrontation between the sisters. As a result, the child stopped respecting and listening to the caregiver as she had her mother's backing.

The caregiver reported that if it was not for the counselling she was getting from their aunt, she would have asked her sister to take her children because she was no longer willing to stay with them. It was after her niece fell pregnant that she managed to talk to her sister about her approach to parenting. She then opened up to her sister that she spoilt the children and they discussed the best way to manage them while they ensure proper discipline. Their other misunderstanding was on the gifts and remittances the migrant sister used to send to the children. The caregiver reported that although it was a good thing for her sister to send money and other gifts, it was important that it is not done excessively. She believed that if the children had excess money, they would misbehave. The caregiver, however, admitted that the confrontations that she had with her sister over the years had affected their relationship and at one point, she was ready to give up.

Both migrants and non-migrant participants reported such confrontations in terms of child discipline. They seemed to disagree on the best ways of training and disciplining the children. Migrants felt that caregivers were too harsh on their children while caregivers on the other hand believe migrant parents were over protective of their children and it became difficult for them to reprimand the children whenever they did something wrong. Such contradicting approaches to parenting were highlighted by both migrants and non- migrant participants in the study. As such, some non-migrant participants reported that they ended up turning a blind eye to the mischievous behaviour of the migrants' children to avoid confrontations with the migrants. As such, the children ended up with ill-behaviour. Smit (2001) reported that children of migrants are characterised by delinquent behaviours, that can be supported by the findings from this study.

Some migrant participants indicated that caregivers did not put maximum effort in caring and nurturing the children, like they did with their biological children. One migrant father reported that all children would be mischievous at some point, but the parent or the guardian should have the upper hand and not allow the children to do as they please,

> All she does is complaining that your son comes home late, he doesn't listen when I talk to him. But what is she doing as a parent, after all she also has children of the same age as mine? Does it mean she does nothing about it too?

In the end, the relations between the migrants and their non-migrant family members they left as caregivers are affected.

In trying to avert such misunderstandings, some non-migrant participants reported that they opt to just not do anything about the behaviour of the migrants' children. They argued that it was better to leave them not reprimanded than to risk clashing with their migrant relatives. However, given their cultural obligations, some of the caregivers reported that it was difficult to just watch their nieces and nephews behave rowdily and yet they do nothing about it. In the Shona culture, the brother or sister's child is like one's own child. It was therefore difficult for most of the caregivers to turn a blind eye to the children's mischief.

Another fallout between migrants and caregivers was reported to emanate from migrant's capacity to remit and the expectations of the caregiver. Although remittances were mostly for consumption smoothing and reducing vulnerability to poverty (Maphosa, 2007; Sikder & Ballis, 2013), some participants reported that they were not enough to cover all the household expenses in the household back home. A non-migrant caregiver in the study who is looking after her son's three children highlighted the importance of remitting, especially if migrants have left their children back home. She admitted that although her son and his wife send remittances, they were never enough to cover their children's needs.

> Yes, they send groceries and some money, but they barely cover the children's school fees. The problem is that when they remit it's more for show so that they can say we send something but what they send is barely enough to provide for the family here.

She indicated that her son and his wife often sent a few groceries and they were insufficient for feeding the family for a whole month. She indicated that she always had to find other means to provide for the children because she was their caregiver. Such reports are in support of Maphosa (2007) who posits that family members left behind expect the migrant to remit once they settle in the host country.

The daughter-in-law on the other hand, gave a different viewpoint. She reported that they always send remittances back to her mother-in-law who was looking after their children. She was also aware that her children were struggling back home. However, she believed that the problem was that her mother-in-law was taking care of many children, and she and her husband, were the only people supporting the family. As such, the remittances that they transferred were not enough to provide for everyone in the household back home.

> We always send groceries, money and clothes back home but the problem is that there are so many people at home. In the end, the groceries do not last. I wanted to take my children to my sister, but my husband refused, saying his mother will look after them but I knew it was not going to be easy because there are other children that she is taking care of.

Discussion

When family members migrate, especially mothers, arrangements are made for someone to provide childcare in the home country. Parents identify individuals they believe are suitable to look after their children during the migration period. Selection of non-migrant family members to engage with in transnational activities is therefore not indiscriminate. Such is highlighted in Mukwembi and Maharaj (2018) who argue that transfer of remittances during the migration period is not random. Migrants transfer remittances to particular individuals in the family. Such has been highlighted in this study as it indicated that familial relations before migration played a major role in selection of caregivers during the migration period. Migrants carefully selected their caregivers by evaluating the strength of social ties they had and the implications of leaving their children with that particular individual.

Migration in the contemporary era is characterised by transnational activities, where migrants remain in touch with their country of origin and work together with non-migrant family members in pursuing different projects and initiatives. It is in such interactions that migrants identify the most relevant individuals to work with. Such interactions are more pronounced when migrants leave their children back home (Maphosa, 2007; Chikanda & Tevera, 2008). Migrant parents are more likely to engage in transnational activities and remit back home than those who migrate with their children.

However, in some instances, relationships between migrant parents and non-migrant family members entrusted their children with are compromised by the constant interaction they have during the migration period. As this study indicated, caregiving during the migration period makes relations between migrant parents and caregivers fragile and can be affected by a number of things. Caregivers want to care for and nurture the migrants' children from their own understanding of parenting, that sometimes is not acceptable by the migrant parents. In the end, caregivers are caught between disciplining and reprimanding the children and risk affecting their relationships with the migrant parents, or leaving the children to behave anyhow and maintain the good relations they had with migrant parents before migration. Such conflicts of interest explain results identified in some studies done on children of migrant parents in different countries. The studies by Safta et al. (2014) and Smit (2001) highlight this idea, as they state that the behaviour and attitudes of some children of migrant parents resemble those of unsupervised children. They highlight that although some migrant parents organise surrogate parents for their children, the children tend to receive minimal care from the caregivers. There seems to be minimal emotional attachment between the surrogate parents and the children. Such findings support Hoang et al. (2015) who argues that care arrangements for children left behind is characterised by negative reports.

Although literature shows that ill behaviour of migrant workers' children is attributed to psychosocial effects of separation from their parents, in some instances it could be caused of tension between migrant parents and non-migrants as indicated in this study. Migrant parents accuse caregivers of ill-treating their children while caregivers argue that they are just training the children to do household chores and reprimand them the same way they do their children. Depending on the age of the children involved, they can realise the tension between their parents and the caregivers and use it to their advantage. Safta et al. (2014) argue that migrant parents often experience conflicting emotions. On the one hand, they want to go home and be with their children, while on the other hand, they want to remain in the host country so that they can provide for their families and invest in the home country. As they choose to stay in the host country, parents compensate for their absence by sending gifts and money to their children back home. Safta et al. (2014) point out that where gifts and money are received by children; such children are likely to become defiant of their caregiver's authority. The results from this study support such arguments, thereby highlighting that it is not entirely psychosocial effects of separation of children and their parents that causes delinquent behaviour.

Conclusions and Lessons Drawn

The study affirms findings from other studies that the volume of migration in the contemporary era has increased and migration is now open for both males and females. Feminisation of migration has more observable effects on the family structure and living arrangements. When children are left in the country of origin, care arrangements are organised and a caregiver is appointed to look after the children during the migration period. Migrants appoint individuals they had good relations with before migration as caregivers. However, their relationships become vulnerable due to different parenting preferences. The study revealed that migrant parents and non-migrant caregivers have different parenting preferences where they fail to agree on the difference between training children house chores and abuse. As such, tension develops between them that in the end negatively affects their relationships.

Given positive contributions of migration to the welfare of families and individuals in the sending countries, it is important that these negative social effects are addressed so that parents can continue working in the host countries knowing that their children are taken care of. It is important that migrant parents and non-migrant parents agree on the general parenting principles, rather than assuming that they are on the same page. If the two parties could agree on how to handle discipline issues, their relationships could be maintained and even get strengthened. Such is important for maintaining familial relationships during the migration period. Relationships during the migration period have the potential to either strengthen or weaken family and kinship social ties. More attention should therefore be placed on childcare given its potential effect on familial relationships during the migration period.

Moreover, it should be noted that the delinquent behaviour of children left behind is not directly a psychological effect of separation from their parents as some is a result of disturbed family relationships. Depending on their age, children of migrant parents can capitalise on the tension between their parents and caregivers to lead a mischievous lifestyle. Assessing the behaviour of children of migrant parents should take into account the familial relations. It is therefore important that efforts to help such children should not focus on them alone, but rather family counselling should be availed to holistically address the problem.

References

Adepoju, A. (2006). Leading issues in international migration in sub-saharan Africa. In *Views on migration in Sub-Saharan Africa: Proceedings of an African Migration Alliance Workshop*. HSRC Press.

Antnam, F. M. (2012). The impact of migration on family left behind. (IZA Discussion Paper No. 6374).

Baldassar, L., Baldock, C., & Wilding, R. (2007). *Families caring across borders: Migration, ageing and transnational caregiving*. Palgrave Macmillan.

Barr, A. (2004). Forging effective new communities: The evolution of civil society in Zimbabwean resttlement villages. *World Development, 32*, 1753–1766.

Chikanda, A. & Tevera, D. (2008). Development impact of international remittances: Some evidence from origin household in Zimbabwe. *Global Development Studies, 5*(3), 273–302.
Cohen, L., Manion, L., & Morrison, K. (2002). *Research methods in education* (6th ed.). Routledge.
Crush, J., Tawodzera, G., Chikanda, A., Ramachandran, S., & Tevera, D. S. (2018). *Migrants in countries in crisis: South Africa case study: The double crisis-mass migration from Zimbabwe and Xenophobic violence in South Africa.* ICMPD.
Crush, J. & Tevera, D. S. (2010). Zimbabwe's exodus: Crisis, mgiration and survival. Cape Town, SAMP
Creswell, J. W. (2009). *Mapping the field of mixed methods research.* Sage.
Gibson, J., McKenzie, D., & Stillman, S. (2011). Impacts of international migration on remaining household members: Omnibus results from a migration lotery program. *The Review of Economics and Statistics, 3*(4): 1297–1318.
Graham, E., & Jordan, L. P. (2011). Migration of the parents and psychological well-being of left behind children in Southeast Asia. *Journal of Marriage and Family, 73*, 763–787. https://doi.org/10.1111/J.1741-3737.2011.00844.X
Demurger, S. (2015). Migration and families left behind. *IZA World of Labor*. https://doi.org/10.15185/Izawol.144l
Dissanayake, P. L., Chandrasekara, N. V., & Jayasundara, D. D. M. (2014). The impact of mother's migration for work abroad on children's education. *The Journal of Science and Technology, 3*(1), 10–120.
Hennink, M., Hutter, I., & Bailey, A. (2010). *Qualitative research methods.* Sage.
Hoang, L. A., Lam, T., Brenda, S. A., & Elspeth, G. (2015). Transnational migration, changing care arrangements and left-behind children's responses in South-East Asia. *Children's Geographies, 13*(3), 263–277.
International Organisation for Migration. (2005). World migration costs and benefits of international migration 2005. *IOM World migration report series.* Downloaded from https://publications.iom.int/system/files/pdf/wmr_2005_3.pdf on 15 March 2017.
International Organisation for Migration. (2010). *Migration and transnationalism: Opportunities and challenges.* IOM.
International Organization for Migration. (2017). *World migration report 2018.* Geneva, IOM.
Jovchelovitch, S., & Bauer, M. (2000). Narrative interviewing. In: Bauer, M. W. G., G. (Ed.) *Qualitative researching with text, image sound.* SAGE Publishers.
Konseiga, A. (2006). Household migration decisions as survival strategy: The case of burkina faso. *Journal of African Economies, 16*, 198–233.
Maphosa, F. (2007). Remittances and development: The impact of migration to South Africa on rural livelihoods in Southern Zimbabwe. *Development Southern Africa, 24*(1), 123–136.
Miller, J., & Glassner, B. (1997). The 'Inside' and the 'Outside': Finding realities in interviews. In: Silverman, D. (Ed.), *Qualitative Research in Psychology*, 99–131–148. Sage.
Mukwembi, T. R., & Maharaj, P. (2018). Understanding relationships and remittance flow during the migration period: Strength of social ties as a factor determining remittance behaviour. *African Human Mobility Review, 4*(3), 1417–1438.
Mu, R., & De Brauw, A. (2013). Migration and young child nutrition: Evidence from Rural China. IZA Discussion Paper No. 7466.
Ratha, D., Mohapatra, S., & Scheja, E. (2011). *Impact of migration on economic and social development: A review of evidence and emerging issues.* The World Bank.
Safta, C. G., Stan, E., Iurea, C., & Suditu, M. (2014). Family destructuring as a results of workforce migration Romanian Realities. *Procedia-Social and Behavioral Sciences, 116*, 2549–2555.
Schmalzbauer, L. (2004). Searching for wages and mothering from Afar: The case of honduran transnational families. *Journal of Marriage and Family, 66*, 1317–1331.
Sikder, M. J. U. & Ballis, P. H. (2013). Remittances and life chances: A study of migrant households in rural Bangladesh. *Migration Development, 2*, 261–285.
Smit, R. (2001). The impact of labor migration on African families in South Africa: yesterday and today. *Journal of Comparative Family Studies, 4*, 533–548.

Smith, A., Lalonde, R. N., & Johnson, S. (2004). Serial migration and its implications for the parent-child relationship: A retrospective analysis of the experiences of the children of caribbean immigrants. *Cultural Diversity and Ethnic Minority Psychology, 10*, 107–122.

Chapter 10
Resilience as a Governance Factor: The Case of Institutional Multiplicity in the Planning of Caledonia, Harare

Tinashe Bobo and Audrey N. Kwangwama

Abstract This study joins a vibrant conversation in the built environment discourse about the planning of peri-urban areas and discussions about institutionalism. It explores the implications of institutional multiplicity to peri-urban planning in Zimbabwe using Caledonia peri-urban development to the east of the City of Harare as a case study. Collectively, the narratives in this study articulate peri-urban dilemmas, such as contestation, because of the existence of multiple institutions and organisations that claim control and custodianship of these areas. Resultantly, peri-urban areas usually lack proper services and infrastructure to support huge populations characterising them and this is a major concern to urban planning since it has a bearing on the achievement of sustainable development. The study engaged document review and case study analysis of Caledonia peri-urban settlement. The results indicate gaps, duplications and confusions associated with institutional multiplicity in peri-urban planning. Hence, there is need for a policy response to address fragmentation of stakeholders' roles in peri-urban areas since they appear orphaned and cursed in terms of planning and development. Institutional interests, pursuits, land rights, management and governance of peri-urban areas have a bearing in peri-urban planning and for the case in study, all these have brought chaos, ill development and illegality in Caledonia. A review of existing planning policies to suit peri-urban planning is critical. The study also recommends for huge financial investments for the provision of bulk infrastructure and public facilities, less reliance on piecemeal planning, collaboration among institutions in land use and development planning.

Introduction

The purpose of this research is to examine the nexus between institutional multiplicity and the planning of peri-urban areas. Institutional multiplicity is defined as the presence of more than one institution claiming control over a piece of land or a geographical location (Di John, 2008). The research has been influenced by the

T. Bobo (✉) · A. N. Kwangwama
Harare, Zimbabwe

I. Chirisa and A. Chigudu (eds.), *Resilience and Sustainability in Urban Africa*, Advances in 21st Century Human Settlements,
https://doi.org/10.1007/978-981-16-3288-4_10

rise of ill-planned and ill-developed peri-urban settlements in Zimbabwe regardless of the existence of various forms of planning, management and governance in these areas. Focus is on Caledonia peri-urban area that represents ill-development and lack of proper peri-urban planning in terms of the spatial dimensions of the settlement and the state and availability of services, such as proper road networks, water and sewer reticulation, power supply and other public services, such as schools and medical facilities, among others.

This research will unearth gaps, duplications and confusions in the planning of Caledonia settlement. It highlights institutional multiplicity as a cancer in the well-being of settlements in terms of peri-urban planning. Hence, the study intends to bring forth recommendations to improve on land use planning so that it coordinates current and future societal needs while minimising conflicts from institutions involved.

Background to the Study

Rapid urban growth in peri-urban areas has plunged many institutions or organisations into peri-urban land conflicts. It has been observed that land availability is often acute in the context of rapid urbanisation and is further worsened by high costs of shelter and servicing that have increased the demand for serviced residential sites that far outwits supply of suitable land. Hence, land becomes scarcer and demand more intense. The regulation of land tenure and land rights in peri-urban areas, has usually been subject to rural land management regimes despite the expansion into them of urban land uses. Such encroachment is often inappropriate or confused (Allen, 2003). Thus, uncertainty over that land tenure regimes are operating, the existence of a variety of different actors and their competing claims to land and lack of administrative clarity and capacity all contribute to the likelihood of land conflicts developing. Globally, land tenure is understood as “the mode by that land is held or owned or the set of relationships among people concerning land or its product” (Payne et al., 2014:416). Accordingly, USAID (2005) defines land tenure as a “bundle of rights” over a particular piece of land. This definition is helpful in identifying the different interests in a given piece of land. Land may have multiple users with specific rights ranging from limited to full use and transfer rights, among others. Therefore, land rights are diverse. Across the globe, these have had differing land use skirmishes in peri-urban areas.

Sub-Saharan Africa has a coexistence of several land tenure systems, such as customary, state, council systems and various indigenous institutions coexist with other forms of social and political organisations exercising control over resources (Amanor & Moyo, 2008). These coexistence dynamics often results in problematic interactions between different institutional forms, especially in sensitive resources, such as land. The Kingdom of Eswatini (former Swaziland), for example, has experienced the effects of institutional multiplicity due to the existence of various institutions in land control and management. According to Goodfellow and Lindermann (2013), this dynamic coexistence often results in problematic interactions between different institutional forms. These problems result in unclear relationships between authorities, such as the local and central government, traditional leaders, politicians and technocrats. This problem has been felt largely by residents in the informal settlement of Moneni that is located in the City of Mazini where traditional leaders and urban government authorities that is, the Mazini City Council and the Ministry of Housing and Urban Development were engaged in contests over land control and authority.

In Zimbabwe's urban areas, such as Harare, the challenge of institutional multiplicity has also included fights between local authorities and traditional leaders but this study is focusing more on the effects of institutional multiplicity to peri-urban planning. This is rooted in the fact that institutional multiplicity in peri-urban areas is usually amongst local authorities and the central government. This scenario has affected land delivery and servicing in peri-urban areas. Peri-urban areas have become areas of interest recently in Zimbabwe's urban landscapes due to their locational advantage to both the affluent and low-income earners. They are characterised by low costs of obtaining un-serviced stands for residential and business purposes. Regardless of high demand for housing space in peri-urban areas, there has been institutional overlap in the planning of these peri-urban areas that has affected their planning and development.

This research focuses on Caledonia farm to the east of Harare that is referred by Mbiba (2017) as an 'organic growth area' that forms a continuation of the existing city (Harare). The area has been under the control of various institutions in terms of planning and development and it rose through the dustbins of informality through socio-political factors that took centre stage after the Fast Track Land Reform Programme (FTLRP) (*ibid.*). Geographically, Caledonia peri-urban area lies within Goromonzi District. However, with peri-urban development menace witnessed in the past, fights emerged over the control of the majestic informal settlement that had overtaken the farm with an approximate 27, 102 people living there by 2012 (*ibid.*). In 2012, the area received a Presidential Proclamation through Statutory Instrument (SI) 119/2012 that declared the incorporation of Caledonia into Harare Municipality.

The Location of Caledonia

See Plate (1).

Literature Review and Theoretical Perspectives

Urbanisation is a global phenomenon that is occurring rapidly in developing countries than in developed countries in the twenty-first century. Topping the list of the most urbanising countries in the world is Eastern and Southern Africa (Nkwae, 2006).

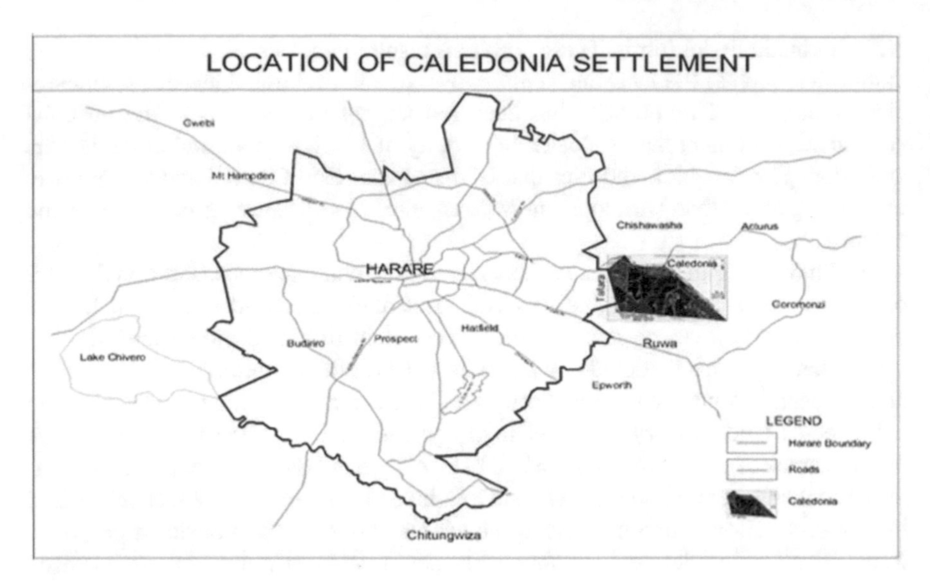

Plate 1 Location of Caledonia (Author 2019)

These parts of the globe have peculiar experiences in terms of urbanisation, originating from the fact that their urban centres were designed and planned for small populations that make them less responsive and resilient. Hence, urban centres are becoming ineffective and inadequate in supplying serviced land for housing to the growing population. Resultantly, these urban dynamics breed diverse informal sector activities in both the urban and peri-urban environments. As cities grow outside their original boundaries into rural areas, there is bound to be conflicting interests in terms of land administration, planning and institutional control of peri-urban areas. This research therefore, tries to interrogate the impacts of having multiple institutions responsible for peri-urban planning and development.

Theoretical Framework

Three theories were adopted in the research in trying to connect peri-urban land, institutions that do planning in peri urban areas and their various interests, pursuits and rights. The theories include the Evolutionary Theory of land rights, land as a Web of Interests and the Institutional Theory. The Evolutionary Theory of land rights was propounded by Platteau (1996) by contending that population pressure in peri-urban areas exerts great pressure on land resources and leads to increased individualisation of land access, increased conflicts between land users and a growing demand for formal property rights. Thus, governments in response to such alarms step in to initiate formal systems of registered individual ownership. The state in this instance will introduce formal registration of land that breaks other forms of land ownership, such as customary land ownership. The theory has proved its practicality throughout Africa, especially in Southern Africa, as evidenced by land disputes and a plurality of interests over land in cities. Therefore, the Evolution of Land Rights fortifies this research.

The research sought an understanding of the relationships among land right holders and the relationships of right holders and the land they have rights over. Hence this explains the phrase “Web of interests”. The proponents of this theory value the relationships between person to person and the relations they have over land. There are different interests for different land right uses. According to Barry (2015) land is tantamount to a web where interests of right uses are intertwined. Central to this theory is land, and interests vary on this land with some having rights to use, some to regulate while others have rights to manage. These variations are explained by (Barry, 2015) to be emanating from a range of customary institutions or local norms and state laws. Important to this Land as Web of Interests theory is that there are different meanings and values attached to land that imply different interests and pursuits on land. A good example is provided by Lund et al. (2006:4) when he argued that, “land in Africa is never just a commodity or a means of subsistence. It has so many other meanings and combines being a factor of production with its role as family or community property, a capital asset and a source of cultural identity and/or citizenship”. This theory resonates well with areas that there are a diversity of stakeholders that this research calls “institutional multiplicity”, different layers of jurisdiction and plural tenure systems, that the end of the day, causes land control conflicts among stakeholders.

This research is hinged on institutional responses to the planning of peri-urban areas. At this point, a synthesis on institutions and the institutional theory will be outlined, specifically answering to the significance of institutions, definitions of institutions, characteristics of institutions, typologies of institutions, and major institutional theories, such as historical, rational-choice and normative. To start with, institution is a slippery term, that means different things to different scholars. An institution can be defined as “the bearer of a set of practices, a structural arrangement and a configuration of rules, that determines what is exemplary behaviour”. These institutions are devised by individuals (micro level) but, in turn, constrain their actions. They are part of the broad social fabric (macro level) but also the conduit through that day-to-day decisions and actions are taken. Hence, Ahmed (2015) argues that institution is a meso-level concept. Institutions evolve as formal or informal. On one hand, formal institutions are normally established and constituted by binding laws, regulations and legal orders that prescribe what may or may not be. On the other hand, informal institutions are constituted by conventions, norms, values and accepted ways of doing things, whether economic, political or social.

Modelling Peri-Urban Land and Planning Issues

Peri-urban land governance in the developing world is shaped by divergent or complementary roles of actors emanating from their authority, power and interests that create a complex relationship affecting land governance processes. This section tries to present the land governance dynamics in peri-urban areas that the research believes has an impediment on the ultimate planning of peri-urban areas and their development. The main focus will then be to identify various gaps, duplications and confusions arising from the roles of different actors, interaction and power relations in peri-urban areas. Scholars have projected that by 2050, more than 66% of the global population will live in urban areas and urban expansion will occur more in Asia, Africa and Latin America. A larger part of these populations will continue to live in a state of illegality in terms of land tenure (Nuhu, 2018). Linked to that population projection, it should be noted that cities mostly in the developing countries will experience high rates of urban expansion where in towns and cities there will be much of vertical expansion while in peri-urban areas, there will be massive horizontal expansion (urban sprawl). Resultantly, cities will and are eating into the boundaries and jurisdictions of small urban and rural centres that are now being turned into peri-urban areas.

According to Adam (2014), peri-urban areas in African cities experience loss of land that has adverse social impacts. In many of these cities, weak urban land governance drives the urban poor to live under the constant threat of eviction (Kedogo et al., 2010; Palmer et al., 2009). Furthermore, people residing in these informal settlements are exposed to various inequalities that range from poor development and lack inclusion into governance and planning institutions. Nuhu (2018) has highlighted that due to the problems associated with poor land governance, attempts have been made to address urban land governance challenges. For instance, in Africa, the state has enjoyed monopoly of power over land but this has been redressed, especially in the post neo-liberal period (Briggs & Mwamfupe, 2000) that saw an increase in the number of international, national and informal actors (Kedogo et al., 2010; Msangi, 2011). The actors are mainly involved in the process of land access in the peri-urban areas of major cities (Gwaleba & Masum, 2018). They have different ideologies and interests on land; hence they exert varying influence over how land is governed in the peri-urban areas. However, it still remains blurred how the interaction of these actors in peri-urban land impacts on the planning and development of peri-urban areas that are mainly illegal and ill-serviced settlements.

There are many actors in the peri-urban land spheres. According to UN-Habitat (2010) the roles of the actors in urban land market do not only cover numerous areas, relations and contexts of activities, but also often overlap; rendering decision-making to be complex. On that basis, peri-urban governance is dependent upon the interaction and power relation among actors with different roles (Katundu et al., 2013). The most outstanding factor is power, that is argued to create exclusion of other actors in the peri-urban land governance. According to Haapanen (2007), exclusion affects the contribution towards improved land governance and the contribution of certain actors in land governance is hindered by these power relations.

Going into the substantive roles of actors in peri-urban land governance, it should be highlighted that central government plays the multifaceted role as a land owner, policy regulator, land administrator and land-use manager and creator of standards and parameters to be followed by local authorities and other actors. On the other hand, government protects its citizens from manipulation by investors, land developers and other interested stakeholders in peri-urban land. Due to these dimensions of peri-urban land, one may argue that peri-urban land is a gold-mine for many that explains the scramble for land and the control of peri-urban lands across the globe. On this note, Nuhu (2018) presents that the urban land market is an arena where urban actors, such as government, private developers, land owners, traditional authorities and others.

The state is responsible for the provision of housing, transport infrastructure and serviced land in peri-urban areas (Mabin et al., 2013). This has been justified by UN-Habitat (2010) that argues that state intervention is necessary because not all urban dwellers can afford to access land. However, there is some confusion about the government actors and their designated responsibilities and roles that create a monopoly of the state over land issues. In some instances, this strengthens state monopoly, especially, when stakeholders are not aware of their mandate and their land rights. Nuhu (2018) provides an example where the state exercises its right to the expropriation of public land where citizens' concerns are not adhered to and the government's decisions are irreversible. Another case scenario is that of power overlaps, where the role of the ministry responsible for land converge with that of local government. In some instances, both ministries responsible for land and local government play similar roles, such as regulation of land uses, allocation of land and controlling development over it. Accordingly, Massio and Norman (2010) argue that in operation and implementation of their duties and roles, in some cases the roles of the government, private developers, land owners, traditional authorities overlap each other, resulting into inefficiency in peri-urban land and planning matters.

Nuhu (2018) refers to the private sector as an influential actor after the government in peri-urban land matters. Through the urban land market, various players in the private sector can access land and are also able to influence most land decisions. Players in the private sector, are driven by the supply and demand of land services. Investors in land related projects and businesses are regarded as part of land-use conflicts in peri-urban areas, with others being the central and local government, communities and local leaders (Kombe, 2010a, 2010b; Msangi, 2011; Makwarimba & Ngowi, 2012). Another group of actors responsible for peri-urban land are the civic society organisations that are at the helm of fighting for citizens' land rights. The organisations are often overwhelmed by political coercion and influence (Haapanen, 2007). From the synthesis above, it is evident that government and its agencies and the private sector, constitute key actors that have the leading role in land acquisition and administration. Nuhu (2018) also highlights that efforts of civic society organisations may be thwarted by the government through its repressive mechanism and regulation of their activities. This identifies well with North's (1990) perception that institutions can constrain actors' sets of incentives and disincentives. This makes actors powerless to undertake community transforming initiatives to avoid conflict with government. On the other category, individual land occupiers and local communities belong to a vulnerable group in society. These are easily exploited by other actors. Mabin et al. (2013) also highlight that the government and sometimes the private landholders can something come into conflict due to rapid urbanisation, frustrating the lives of peri-urban dwellers. This is also coupled by the fact that land occupiers in peri-urban areas themselves are poorly mobilised, lack financial resources and thus do not influence most land governance decisions. This factors in the challenge of local communities' participation in peri-urban land governance and planning.

Emerging Planning and Land Issues in Peri-Urban Areas in Zimbabwe

Peri-urban areas have not only become the epicentre of political struggles in recent years, but have also become a major concern ''in urban planning and land management in large cities in Zimbabwe''. Just like any other developing country in the world and being a typical African nation, Zimbabwe is experiencing high rates of urbanisation. Urbanisation usually imposes unanimated pressures on existing artificial ecosystem structures and services that explains the huge demand for housing in urban and peri-urban areas. Various sources of literature (Ahmed, 2015; Allen, 2003; Barry, 2015; Bloch, 2015; Briggs & Mwamfupe, 2000; De Soto, 2000; Di John, 2008; Gwaleba & Masum, 2018) have argued that people seek homes in peri-urban areas driven by the tradition that places weight in rural home ownership regardless of having an urban home. This has since changed due to the dynamics in the housing sector since many low-income people are finding it cheaper in establishing their homes in peri-urban areas and commuting to town for work. Dynamics in the demand for housing in peri-urban areas have attracted illicit land dealers who engage in the business of buying cheap and selling high.

This has been elaborated by De Soto (2000), in his book entitled the, 'Mystery of Capitalism' where he talks of the *dead* capital that lies in property, especially from the poor people's perspective. Although De Soto (2000) believes that most developing countries have *dead* capital lying idle, Zimbabwe has identified such capital in a different manner such that urban and peri-urban areas now support a thriving and lucrative informal land/housing market. On this same stance, the control of urban land has become intensely political. Despite building and urban land management and regulation in Zimbabwe being highly perceived, there are ambiguities in the peri-urban strata. The continued selling and parcelling of land in peri-urban areas of cities, such as Harare, have created contested spaces due to institutional multiplicity in that various authorities claim responsibility in certain aspects and some in other aspects. At the same time, the traditional systems of land governance have not given up on their so called "inheritance", hence, they are alive and well in these areas.

Tanzania Case Study

Tanzania has recorded a 5.3% growth rate of its urban population between 2002 and 2012. This period witnessed urban population growth of around 8.7 million people increasing the urban share of the total population from 22 to 29% (Hugh, 2014). Statistics have highlighted high population growth rates and Tanzania's urban population is expected to increase by 61.5 million between 2010 and 2050. Regardless of the current levels of urbanisation and the expected increase, Tanzania has seen the dismantling of local government through attempts to direct urban planning through central government organs. This has led to an undersupply of well-planned areas and formal housing. Provisions for local government exist but scholars, such as Owens (2014), regard them as seriously limited in their capacity to tax and spend that means struggling also to provide services to deal with the influx of migrants.

Centralised control of land in Tanzania led to the creation of laws, such as the National Housing Corporation (NHC) that carried the mandate of providing housing for all households in urban areas. However, NCH has eventually failed to do so. This led to the creation of self-help initiatives and the rise of informal channels to housing delivery. From the above, land bureaucracy in Tanzania's government has meant that land governance and administration have been struggling to perform well. The main reasons outlined by Kironde (2006) and Midheme (2007) have to do with unclear institutional mandates for planning between different levels of government, lack of resourcing for the projects to regularise land, slow land administration procedures and unrealistically high standards for planned settlements, such as minimum areas for plots. As a result of these institutional aspects and others, local governments have failed to meet housing needs and creation of planned areas for human habitation. This has spiralled to a 40–80% informal settlement coverage of all built up areas in Tanzania (Lupala, 2015) and a situation where many households in all major cities live in unplanned areas (Kyessi & Sekiete, 2014).

Planning and land governance in Tanzania as highlighted above have impinged peri-urban areas. This has been possible through less attention that has been given to peri-urban areas in the policies of urban development and spatial planning. The implications have been presented to be the following:

- Government decisions on municipal boundaries extending into rural areas and designation of rural to urban have impacted on how peri-urban land is governed and resources that exist to support land administration and planning.
- There has been virtually low service provision in peri-urban areas.
- Incentives within the land planning and registration system have led peri-urban areas to be first ignored and then later targeted.

Impacts of urban expansion in Dar es Salaam have meant that attention given from the government's side remained slim, especially in land registration. In addition, some major initiatives attempting to deliver planned settlements at large scale have run into problems because of conflicts over authority of planning and decision-making and questions of how to deal with existing claims to land, including paying compensation to land owners when land needs to be acquired. As such, Kombe (2005) has argued that the development of peri-urban areas around smaller cities and towns happens without the input or oversight from city planners, and landholders do not perceive it in their interests to interact with formal land registration processes. From the Tanzanian experience, it has been substantiated that urbanisation is alive and well in the African region and that it is not a smooth process, but rather a complex one with a host of questions. As urban areas expand, questions on how to manage land in the peri-urban areas arise. According to Locke and Henley (2016) touches on the changing spheres of influence/governance, land administration and planning.

The Kingdom of Eswatini (Formerly Swaziland)

In the Kingdom of Eswatini formerly Swaziland, there is a problem of interface of traditional and urban institutions. The king exercises executive, legislative and judicial powers and the country is ruled through the Tinkhundla Political System that permits the subdivision of the country into 55 constituencies. The constituencies are comprised of 360 chiefdoms that operate as administrative centres. Most migrants settle in unplanned or un-serviced areas. For this reason, the nature of growth in the main cities of the Kingdom of Eswatini, such as Manzini and Mbabane, have been largely informal.

The Kingdom of Eswatini has various legislations or pieces of the law that govern the work of municipalities and how land is earmarked for various uses. The Urban Government Act No 8 of 1969 mandates the minister responsible for housing and urban development to declare an area a municipality and even alter boundaries of existing municipalities. The Kingdom of Eswatini National Physical Development Plan of 1996 dictates that once the areas are gazetted as urban, the traditional authorities cede their jurisdiction over them. This kind of relationship between municipalities and traditional land control systems has created an unhealthy relationship between traditional leaders and urban authorities. There are two factors that have impinged on the relationship that are the fact that traditional leaders often refuse to cede their authority in their previous jurisdictional areas and the fact that urban legislation fails to define and incorporate traditional authority structures into urban land management structures, instead considering them as an impediment to urban development.

What is more interesting in the case of the Kingdom of Eswatini is that the country's legal system allows for the co-existence of statutory and customary laws and permits the monarchy to exercise absolute control over the nation. Laws clash, especially in peri-urban areas, where urban authorities carry out land distribution and development projects in accordance with statutory law, despite the custom of chiefs and traditional councils to decide on land distribution. This shows institutional multiplicity in land governance that is popular in Sub-Saharan Africa. In the Kingdom of Eswatini, informal settlements have become spaces of confrontation between traditional and urban authorities who disagree on land administration. These peri-urban areas have turned into areas of contestation due to lack of clarity arising from the fact that urban expansion has led to the incorporation of rural land into urban areas.

Institutional multiplicity has affected the governance of peri-urban lands through the lambasting of urban government officials end the disturbance of the planning exercises. During community meetings, supporters of traditional authorities dismissed the city council's involvement in the area. According to Simelane (2016), urban government officials had to bear the wrath of residents who did not want the city council to have any authority in their areas. Local supporters of the traditional leaders have tried to keep city council officials out of Moneni and some resorted to smashing the Planning and Community Development Department' vehicles with stones, insisting that the department should not inspect the area. The city council has also tried to stop the construction of new structures in Moneni but some sections of the public have defied notices from building inspectors to stop construction. The new structures witnessed in the Moneni informal settlement were mainly modern houses.

Research Methodology

This study engaged in document review and case study analysis of Caledonia informal settlement and looking into cases from other countries across the globe, such as Tanzania and The Kingdom of Eswatini (formerly Swaziland). The choice of the case study was based on the notoriety of the Caledonia settlement in Zimbabwe where formality, informality, illegal and legal land development coerce and the presence of varied institutions responsible for that area. Secondary data was collected through documents, such as scholarly articles and policy documents. The study also benefited from primary data collection that involved interviews carried out with officials from Goromonzi Rural District Council, City of Harare, Urban Development Cooperation (UDCORP), Department of Spatial Planning and Development, Housing Cooperatives and Residents of Caledonia. Subsequently, data generated from the document review was analysed through thematic and narrative analysis. Primary data was analyses through the Statistical Package of Social Sciences (SPSS). Themes were generated from the data to give meaning to the topic under study.

Caledonia consist of about 29,000 households that was the sampling frame for this study. The 29,000 households are spread over about 21 phases that are treated as clusters in this study. This was a large number to consider in doing cross-sectional studies, hence a small sample was selected to represent the wider peri-urban area. 10 phases are selected for the study. Information is gathered from this sample and for the purposes of this research, the respondents are outlined in Table 1.

Research Findings and Analysis

The Planning Process versus the Development of Caledonia Settlement

Table 1 Sample composition totals

Respondents	Sample size qualitative
Harare city council	1 Participant
Goromonzi rural district council	1 Participant
Department of spatial planning and development	1 Participant
Urban development corporation (UDCORP)	1 Participant
Housing cooperatives	1 Participant
	5 Key Informants

Source Author (2019)

Peri-urban planning is a technical process that is concerned with the use of land and the design of the peri-urban environment. These developments do not happen automatically but require control since they happen on land that is a scarce resource or commodity. Also, the erection of buildings and structures over land must be done in an orderly fashion. Thus, planning in Zimbabwe is regulated through the Regional, Town and Country Planning Act (RTCPA) Chapter 29:12 of 1996 as revised in conjunction with other allied acts. The RTCPA outlines both procedural and substantive issues in the planning of human settlements and it promotes order, safety, health and amenity, among other principles. If better settlements are to be created, then the procedures for planning or rather land development should be well followed. However, this study argues that Caledonia as "an orphaned" settlement resembles a great deal of failure in the planning and land development processes of peri-urban areas. At this point, the planning process will be outlined in sync with the information obtained during data gathering from key informants from various institutions.

Land is not developed from scratch, there are procedures in planning and in land development processes. An official from Harare City Council highlighted in an interview with the researcher that,

> The process of land development starts from the preparation of master plans, local development plans and layout plans.[1]

These plans are known as forward plans, in that they point to the future of an area in terms of planning and development. They are provided for through the RTCP Act Chapters 14 and 18.

Scholars, such as Duke (2017), have argued that "just as a building requires thorough, detailed architectural plans to affect good construction, so does a municipality need good plans to affect good development. Hence, master and local plans outline policies with regards to the development of places and land use zones in particular areas. However, key informants provided that Caledonia grew to an outstanding settlement comprising of 21 phases with an approximate of 29 000 households

[1] Interview, Harare City Council Official, August 2019.

without master and local plans provision and proper layout planning. An official from UDCORP said that,

> Phases 1-3 were the only phases with approved layout plans that were prepared by PlanAfric Consultancy, and from phases 4-20 there were no approved layout plans even though they were fully developed (informally).[2]

The official went onto say that for Phases 4–20 of Caledonia, layouts were prepared by a consultant and handed over to housing cooperatives for implementation without approvals. The implication of these developments in Caledonia meant the majority of constructions were done without formal and informed town planning advice on stand sizes, building lines, size of permissible dwellings, maximum heights of buildings, floor area factors and many other applicable details. Lack of town planning influence has been seen to be the cause of haphazard developments and ill development in Caledonia.

The Department of Spatial Planning and Development (DSPD) is an important stakeholder in the planning of peri-urban areas through the preparation, supervision, approval and implementation of master, local and layout plans. However, since Caledonia did not have an operative master, local and layout plan when people started moving into the area, the Department of Spatial Planning and Development was involved in preliminary planning to house people forced out of Churu and Porta farm. According to an official from the DSPD,

> We took part in the preparation of a quick layout for quick habitation in Phases 1-3 of Caledonia to pave way for people coming in from Churu and Porta farms.

Resultantly phases 1–3 of Caledonia highlight some form of order and development than phases 4–20 that were largely developed from an unapproved layout plan.

In line with layout planning, the RTCP Act provides for processes of subdividing and consolidating land. This is provided in Sections 39–44 in the RTCP Act and it clearly states that a piece of land can be subdivided or consolidated to make two or more stands for the purposes of development after having obtained a permit to do so. However, for Caledonia it was different because by the time Goromonzi Rural District Council realised its mandate according to law, Caledonia farm was already in pieces. An official from UDCORP highlighted that land squabbles due to illegal/informal subdivisions and allocations were so high that they attracted the Minister's attention and a commission of enquiry led by one town planning consultant established that there were 130 housing cooperatives operating illegally and parcelling out land for urban development. This deviates from town planning laws that state that local planning authorities should be responsible for such kind of planning. The official from UDCORP went on to highlight that the District Administrator (DA) for Tafara (Harare City) was implicated in land deals and land-allocations.[3] This shows a major gap in planning and land parcelling, allocations and the ultimate development of the Caledonia settlement.

---

[2] Interview, UDCORP Official, August 2019.

[3] Interview with UDCORP Official, August 2019.

In proper land development processes, when a layout has been approved, it means the area being planned for is ready for a title/cadastral survey. Surveyors carry out the cadastral survey before services are provided. These surveys show the demarcations and boundaries of each piece of land and also link the already developed land outside the planning area with the layout of the area at hand. When surveys are completed, they are examined and approved by the Surveyor General. All this is supposed to happen before an area is occupied since surveys also include the pegging that defines the size, location and dimensions of a particular stand. However, Caledonia was an exception because it was invaded without any form of surveys such that when UDCORP was sworn in as the project manager for Caledonia they started mapping the settlement and carrying out surveys with MACDOH Planning Consultancy[3].

Efforts by UDCORP as the project manager failed to transform Caledonia as they only managed to develop 4.02 km of tarred roads out of the required ± 20,000 kms and 8 bridges. Other developments included the drilling of 22 boreholes against a household's population of 29,000. This highlights a major discrepancy in service provision as demand outstripped supply and amazingly one borehole was supposed to service 1 310 households per day. During the study it was found out that only 10 boreholes out of the 22 provided by UDCORP were still working and the majority were facing breakages and exhausted water yield. These efforts were redeeming in nature in an unserviced settlement but, however, the UDCORP official highlighted that,

> The efforts were hindered by failure by beneficiaries to pay monthly subscriptions of $50 per month.[4]

Certificates of Compliance are issued after the attainment of the required conditions to development. At this point building plans are submitted to the responsible local planning authority for approval and they become ready for actual development. Building plans are either for dwellings, commercial or any other buildings. Construction works are subject to local authority inspections. However, the development of the majority of Caledonia happened without compliance and chaos became the order of the day. The Commission of Enquiry led by a town planning consultant reported that before any meaningful planning processes were employed in Caledonia, the settlement had grown even to the extent of spilling over into land informally designated for public facilities, such as schools and churches, among others.

These massive developments also were not subject to building inspections in Caledonia where building standards and provisions of local plans are checked upon before development is allowed on a particular stand. Resultantly, there was nothing illegal in Caledonia, everyone was an authority and development happened as land barons, housing cooperatives and political figures wished. In other words, disorder became the order of the day. Certificates of Compliance are requisite documents for the Deeds Registry to process title deeds. They prove that a developer has complied with the subdivision conditions from the local authority. A tittle deed can be defined

[4] Interview, UDCORP Official, August 2019.

as a legal document given to the owner of land by the Registrar of Deeds. In Caledonia all the above never materialised.

A certificate of occupancy is issued when one satisfies conditions stipulated by the local authority. This is proof that a house has been built under the local authority's by-laws and is fit for human occupation. In formal settings, this process was prerequisite to the occupancy of any land designated for residential purposes. However, the Parallel Development Policy introduced by government through the Zimbabwe National Housing Delivery Program (ZNHDP) gave a leeway to chaotic urban developments across Zimbabwe where housing developments on individual stands could be carried out whilst roads, sewer and water infrastructure are being developed. This can also be tied to the developments that happened in Caledonia where a blind eye was given for people occupying the land in the name of parallel development.

Intervention and the Development of Institutional Multiplicity in Caledonia

Caledonia was born with a birth right of chaos, poor sanitation and irregular settlement patterns. With such settlement development discrepancies at hand, Caledonia has technically digressed into a notorious informal settlement. There are no roads, electricity, clinics, schools, water and sewer reticulation facilities and other essential social amenities for proper human habitation. Out of the 21 phases in Caledonia settlement, only three phases have received some form of meaningful planning or regularisation. Some of the residents' respondents reported that.

> Some of us have started staying in Caledonia around 2000 and no services have been provided to improve our living conditions.

Lack of services and the scramble for cheap land for housing caused a public outcry that brought government intervention through the then Minister of Local Government, Public Works and National Housing (MLGPNH), (I. Chombo). Minister Chombo received a complaint in 2011 on land squabbles in Caledonia. An official from the Urban Development Cooperation (UDCORP) indicated that the former Minister Chombo; on hearing the public outcry in Caledonia, set up a team to investigate on the land squabbles that was led by Percy Toriro, a town planning consultant. The independent Commission of Enquiry was tasked to research and report on the complaints received by the ministry on land related issues. This resulted in a report documenting findings that proved that the level of chaos in Caledonia had gone out of hand. The following were some of the striking results of the report served to the Minister of local Government, Public Works and National Housing by the Commission:

Box 1: Key Findings of the Commission of Enquiry set over Caledonia Land Issues

- There were 130 housing cooperatives operating in Caledonia, these were doted in 21 phases that constitute the Caledonia peri-urban area.
- The report implicated the District Administrator for Tafara from the City of Harare as responsible for land-allocations in Caledonia.
- Approximately 50 million United States dollars had been lost to housing cooperatives but there was no development whatsoever.
- There was no coordination between various institutions that were active in the parcelling and development of Caledonia.
- 1–3 phases only had approved plans out of 21 phases, the plans were done by PlanAfric consultant
- Layout plans for Phases 4–21 of Caledonia were being implemented (stands allocations and non-title surveys) by housing cooperatives without approvals from any local authority or the Department of Spatial Planning and Development.
- Stands earmarked for institutions, such as schools, churches, crèches and buffers, among others in the unapproved layouts were being invaded. Some of the developments were being done in buffer zones and some even below high voltage ZESA pylons.

Source Interview, UDCORP Official, August 2019.

The Genesis of Institutional Multiplicity in Caledonia

The findings of the Commission paved way for a promising period where some form of coordination was introduced in the affairs of Caledonia amongst various stakeholders. The commission recommended for a project manager. It is on this background that UDCORP was appointed to spearhead the planning and development of the Caledonia settlement. According to a UDCORP official, the appointment of UDCORP was influenced by their track record in projects they did in Manyame and in Marondera.[5] In addition to recommending for a regulatory institution, the Commission of Enquiry also recommended that the regulatory institution should not work alone. This implied that other stakeholders had to be roped in to complement the mandate of UDCORP. This led to the Government of Zimbabwe giving a proclamation in the Government Gazette in 2012 that Caledonia had been transferred from Goromonzi Rural District Council to Harare City Council. This according to this research signalled the genesis of institutional multiplicity in the planning of peri-urban Caledonia that this study underpins to the prevailing chaos.

[5] Interview, UDCORP Official, August 2019.

UDCORP is a parastatal established by Government in 1986 to spearhead urban development. However, over the years, it became incapacitated to handle urban development projects. An official from UDCORP highlighted that the appointment of UDCROP to do project management in Caledonia was premature.

> Government should have re-organised it first to improve its capacity. Its enabling Act should also have been reviewed to give it more enabling powers.[6]

This highlights that the capacity of UDCORP was far below the project management tasks it had been given to regularise informal settlements in Zimbabwe including Caledonia. UDCORP lacked capacity in terms of the adequate financial resources to plan for 29,000 households including the provision of services, such as water, sewer and road facilities. The subscriptions that were being paid on monthly basis by the residents of Caledonia could not cover for the developments required. The UDCORP official also highlighted that they used almost 1.6 million US dollars for surveying only in Caledonia. This shows the fundamental need for huge capital injections from government and private sector through public–private partnerships.

On another dimension, UDCORP as a parastatal lacked the adequate technical staff to deal with the technical demands of Caledonia chaos. For example, UDCORP did not have adequate planners and engineers to do planning and infrastructure development for an area with a population of more than 100 000 people. All these challenges had a bearing on the success of the appointment of UDCORP as the project manager for Caledonia. In another issue raised by the UDCORP official, the project manager did not have powers to enforce the payments of development levies and as such, they relied on goodwill that is a rare animal in land development and housing provision due to the public good opinion associated with housing.

Resultantly, Caledonia became the haven for the following institutions (Figure), claiming various roles and responsibilities and resultantly failing to bring about proper peri-urban planning and development (Fig. 1).

Figure above is a praxis of the development of institutional multiplicity throughout the history of Caledonia settlement. There is a highlight of institutions claiming various roles over the planning, control, development and management of the peri-urban settlement of Caledonia. These include interests, pursuits, management and governance of the peri-urban area. among other.

Gaps, Duplications and Confusion in the Planning of Caledonia in Light of Institutional Multiplicity

The study has highlighted that the planning of Caledonia settlement is characterised by gaps, duplications and confusions. These are the major effects of how planning is being done in Caledonia and this study further maintains that institutional multiplicity

[6] Interview, UDCORP Official, August 2019.

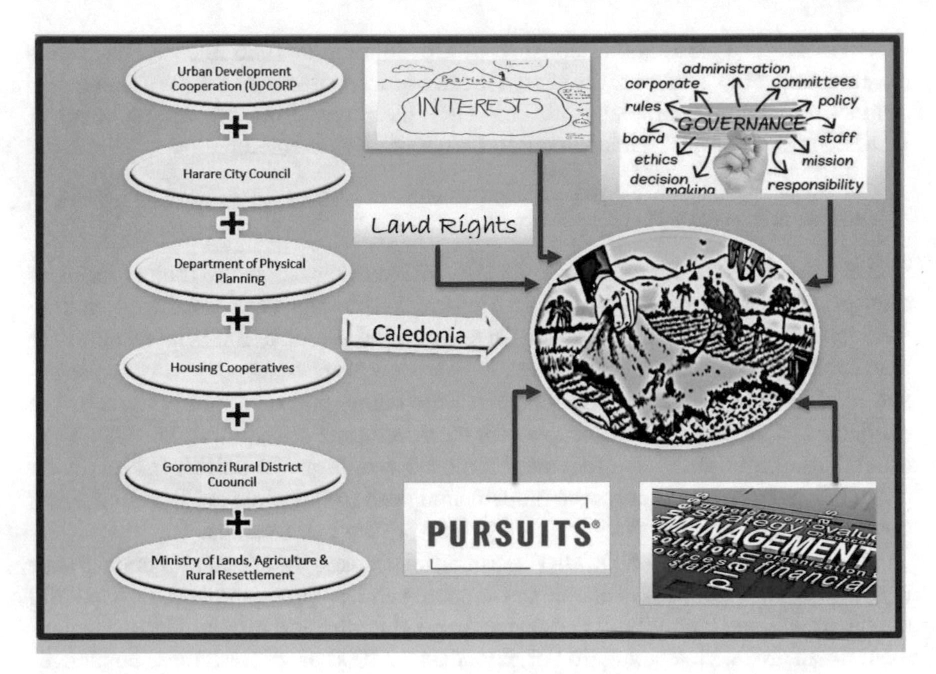

Fig. 1 Caledonia institutional multiplicity. *Source* Author 10 (2019)

is the contributing factor to the current state of Caledonia. Gaps, duplications and confusion can be defined as follows:

- Gaps are considered as the missing pieces in the sustainable planning of human settlements.
- Duplications refers to a situation in that one thing has the same purpose of effect as another and is therefore not necessary. It is also the process of making an exact copy of something.
- Confusion is a state of uncertainty about what is happening, intended or required. This situation entails lack of clarity in planning and land governance of Caledonia (Table 2).

Legislative Confusions in the Planning of Caledonia

In 2012, the Government of Zimbabwe through the Presidential Proclamations 1 and 2 transferred Caledonia from Goromonzi Rural District Council to City of Harare. The reasons behind this move are varied across disciplines, affiliation and viewpoints. An interview with an official from Goromonzi Rural District Council indicated that Caledonia was transferred to Harare in 2012 after the Ministry of Local Government, Public Works and National Housing had realised that Goromonzi did not have

Table 2 Gaps, duplications and confusions in the planning of Caledonia

Gaps	Duplications	Confusion
• No master/local plan for Caledonia (land use provision, unorderly development)	• Layout planning (done by UDCORP and planning consultants (MACDOH Planning Consultancy) instead of City of Harare)	• Appointment of UDCORP (political)
• Piecemeal planning (quick layout and survey)	• Service provision (roads, water and sewer reticulation)	• Control of Caledonia • Administration of Caledonia e.g. Councillor for Caledonia sitting in City of Harare council chambers
• No proper coordination between parties (e.g. Goromonzi versus City of Harare)	• Approval of plans (e.g. bridges were approved by the Ministry of Local Government, Public Works and National Housing whereas City of Harare had been made the Local Planning Authority)	• Banning of Housing Cooperatives by the Ministry of Local Government, Public Works and National Housing whilst they were regulated by the Ministry of Small to Medium Enterprises
• No enforcement (Development Control)	• Site planning done by UDCORP and Plan approval done at City of Harare	• Extensive housing cooperative membership to deal with payments and the planning of the area
• Key players, such as UDCORP, did not have power to enforce the payment of development fees		• Development levy payment method (paid through housing cooperatives)
• UDCORP Developed 4.2 km tarred road and about 20,000 km gravel roads without designs and/or approvals	• Double allocations of stands	• Double allocations of stands
• Settlement form (geography) in relation to sewer reticulation (Caledonia was fit for low/ultra-low density)		• Residents to pay for site plans to UDCORP and then submit building plans for approval to City of Harare
• Parallel development regulations (ZNHDP of 2005)		
• Lack of public participation on the transfer of Caledonia from one local authority to another		• Reversal of Government Proclamation on Caledonia made through Statutory Instrument SI 119/2012 with Statutory instrument 141 gazetted proclamation 3 of 2018

Source Author, 2019

adequate resources and services to provide services and develop Caledonia settlement. Thus, Harare was seen as the only local authority that could provide services to Caledonia. As this came from above, Goromonzi Rural District Council had no option than to let Caledonia be taken by Harare. However, an official from UDCORP argued that the transfer of Caledonia to the City of Harare was a complex and well calculated move by the ZANU-PF led government. Although Caledonia was transferred to Harare, it should be noted that the minister went on to recommend UDCORP as the project manager for Caledonia development. According to the UDCORP official, UDCORP was used;

> as a political tool to paint black on City of Harare, as City of Harare was supposed to be responsible for the planning and running of affairs of Caledonia settlement.[7]

However, the proclamation was operative up until 2018 when what this study terms "legislative confusion" struck the ZANU-PF led government that then stripped Harare of its roles and responsibilities over Caledonia by transferring Caledonia back to Goromonzi Rural District Council. Statutory Instrument 141 Gazetted Proclamation 3 of 2018 that was signed by the President on the 15th of July and gazetted in a Gazette Extraordinary on 27th July just three days ahead of polling day on the 30th July harmonised election. This proclamation was received with mixed feelings with some calling it a "clumsy and defectively drafted proclamation". A general look on the turning of events also highlights some major political connotations, especially of vote-buying gimmick. An official from Goromonzi Rural District Council highlighted that,

> The 2012 proclamation to transfer Caledonia to Harare was not complete and that is why Caledonia should revert back to Goromonzi.[8]

This can also be substantiated by the fact that when Caledonia was transferred to Harare in 2012, there was no hand overtake over from one local authority to the other. The implication was such that Caledonia automatically and suddenly became an orphaned place. The transfer of Caledonia to Harare was done under the speculation that Harare as a major urban local planning authority would be able to deal with the challenges facing Caledonia settlement. However, this study maintains that the transfer of Caledonia to Harare was a political move by the ZANU-PF led government to boost political support in the 2013 National Harmonised Elections. This move was necessitated by the experience that ZANU-PF was losing to MDC in urban areas and, as such, incorporating Caledonia to Harare would neutralise an already MDC-identified support base. This has also been substantiated by the fact that Caledonia's population amounts to a quarter of Harare's urban population.

[7] Interview, UDCORP Official, August 2019.

[8] Interview, UDCORP Official, August 2019.

Conclusion and Recommendations

The study confirmed in terms of literature and the findings obtained in the study area, that urban and peri-urban centres are becoming ineffective and inadequate in supplying serviced land for housing and other public facilities to their growing populations. Nkwae (2006) argued that "as a result of the failure to cater for serviced land for housing and public facilities, informal sector activities in both the urban and peri-urban areas are established". Informal activities are seen as the easiest ways to provide social facilities because they do not follow any set of rules. Also, allowing Caledonia to be passed from 'pillar to post' that is giving overlapping powers to different institutions worsened the fate of Caledonia that by 2012 was suffering already from the activities of housing cooperatives and notorious land barons in light of the failure of Goromonzi RDC to plan for Caledonia.

As such, Goromonzi RDC, Housing Cooperatives, State Lands and Department of Spatial Planning and Development were joined by the City of Harare and UDCORP all targeting to make a mark on Caledonia. This unfortunate condition is termed "institutional multiplicity" by Di John (2008:1) where multiple claims of governance and where other actors than the state engage with the provision of basic services. Caledonia was shoved into a bag of both formal and informal institutions where laws, regulations and standards intermingled with norms and values it is also a place where informality was accepted. Instead of working together, these institutions did more harm than good to the settlement. Since 2012, Caledonia became an informal urban–rural jungle devoid of either amenity yet there are many institutions responsible for its planning and development. On this basis Woltjer (2014) argues that institutional dimensions in peri-urban development brings in the challenge of lack of mechanisms for inter-regional coordination and inter-sectoral integration.

The research recommends for the following:

- Review of existing planning policies and legislation, n such as the Regional, Town and Country Planning Act Chapter (29:12 of 1996) as revised to adequately cover peri-urban planning processes.
- Preparation of master and local plans that direct development and enable development control to be carried out so that order, safety, health and amenity, among others.
- Coordination and collaboration of institutions responsible for the planning and development of Caledonia.
- Extensive research in the planning of peri-urban areas from both the academic and the practical world.
- Huge financial investments from the private and public sectors in peri-urban areas. It was highlighted that Caledonia required approximately $107,640,000 USD for development. Hence there is need for the government and private sector to pool in resources for the planning and development of human settlements in Zimbabwe.
- To kick-start the planning and development of Caledonia, there is need for peace-building and conflict resolution initiatives to lessen tension and mistrust amid institutions, such as Goromonzi Rural District Council, Harare City Council,

Department of Spatial Planning and Development, Housing Cooperatives and Caledonia residents, among others.

References

Adam, A. G. (2014). *Peri-urban land tenure in ethiopia.* Doctoral Thesis in Real Estate Planning and Land Law Real Estate Planning and Land Law Department of Real Estate and Construction Management School of Architecture and The Built Environment Royal Institute of Technology (KTH) Stockholm, Sweden.

Ahmed, T. (2015). *Institutions and institutional theory*. Dhaka University.

Allen, A. (2003). Environmental planning and management of the peri-urban interface: Perspectives onon an emerging field. *Environment and Urbanization, 15*(1), 135–148.

Amanor, K. S., & Moyo, S. (2008). *Land and sustainable development in Africa.* London: Zed Books Ltd.

Barry, M. (2015). *Property theory, metaphors and the continuum of land rights.* Nairobi: Un-Habitat.

Bloch, R. (2015). Africa's new suburbs. In P. Hamel & R. Keil (Eds.), *Suburban governance: A global view* (pp. 253–277). University of Toronto Press.

Briggs, T., & Mwamfupe, D. (2000). The changing nature of the peri-urban zone in Africa: Evidence from Dar Es Salaam. *Scottish Geographical Journal, 115*(4), 269–282.

De Soto, H. (2000). *The mystery of capital: Why capitalism triumphs in the west and fails everywhere else.* Civitas Books.

Di John, J. (2008). Why is the tax system so ineffective and regressive in latin America? *Development Viewpoint, 5*, 1.

Duke, A. (2017). *A town well planned*. New Jersey: Strong Towns.

Gwaleba, M. J., & Masum, F. (2018). Participation of informal settlers in participatory land use planning project in pursuit of tenure security. *Urban Forum, 29*(2), 169–184.

Haapanen, T. (2007). Civil sSociety in Tanzania, KEPA Working Papers Number 19. Helsinki, Finland: KEPA.

Hugh, W. S. (2014) Population growth, internal migration and urbanisation in Tanzania 1967–2012: A census based regional analysis. International Growth Centre (IGC).

Katundu, M. A, Innocent, M. A. M, & Mteti, S. H. (2013). *Nature and magnitude of land acquisitions in Tanzania: Analysing role of different actors, key trends and drivers in land acquisitions.* A Paper Presented at the International Conference On "The Political Economy Of Agricultural Policy In Africa" Held at the Roodevallei Hotel, Pretoria, South Africa, From 18–20th March 2013: South Africa, Future Agricultures Consortium.

Kedogo, J., Sandholz, S., & Hamhaber, J. (2010). Good urban governance, actors' relations and paradigms: Lessons from Nairobi, Kenya, and Recife, Brazil, Good Urban Governance in Nairobi and Recife. In *46th Isocarp Congress 2010.* Nairobi: Isocarp.

Kironde, J. L. (2006). The regulatory framework, unplanned development and urban poverty: Findings from Dar Es Salaam, Tanzania. *Land Use Policy, 23*(4), 460–472

Kombe, W. J. (2005). Land use dynamics in peri-urban areas and their implications on the urban growth and form: The case of Dar Es Salaam, Tanzania. *Habitat International, 29*(1),113–135.

Kombe, W. (2010a). Land conflicts in Dar Es Salaam: Who gains? Who loses? *Cities and Fragile States Working Paper, 82.*

Kombe, W. (2010). Land acquisition for public use emerging conflicts and their socio-political implications. *International Journal of Urban Sustainable Development,* 2(1–2), 45–63.

Kyessi, A. G., & Sekiete, T. (2014). Formalising property rights in informal settlements and its implications on poverty reduction: The case of Dar Es Salaam, Tanzania.

Locke, A., & Henley, G. (2016). In *Urbanisation land and property rights.* Shaping Policy for Development Odi.Org

Lund, C., Odgaard, R., & Sjaastard, E. (2006). Land rights and land conflicts in Africa: A review of issues and experiences. Danish Institute for International Studies.
Lupala, J. M. (2015). Urban governance in the changing economic and political landscapes: A comparative analysis of major urban centres of Tanzania. *Current Urban Studies, 3*(02), 147.
Mabin, A., Butcher, S., & Bloch, R. (2013). Peripheries, suburbanisms and change in Sub-Saharan African Cities. Social Dynamics. *A Journal of African Studies, 39*(2), 167–190.
Makwarimba, M., & Ngowi, P. (2012). *Making land investment work for Tanzania: Scoping assessment for multi-stakeholder dialogue initiative.* Paper Presented at The Tanzania Natural Resource Forum.
Mbiba, B. (2017). Missing urbanisation in Zimbabwe on the periphery. Infrastructure and Cities for Economic Development. Africa Research Institute.
Msangi, D. E. (2011). *Land acquisition for urban expansion: Process and impacts on livelihoods of peri-urban households. Doctorial Licentiate Thesis.* Swedish University of Agricultural Sciences.
Midheme, E. (2007). *State versus community-led land tenure regularization in Tanzania. M. Sc Dissertation.* International Institute for Geo-Information Science and Earth Observation, (Itc), Enschede, The Netherlands. Available Online http://www.itc.nl/library/papers_2007/msc/upla/midheme.Pdf Accessed on 13 July 2019.
Nkwae, B. (2006). Conceptual framework for modelling and analysing peri-urban land problems in Southern Africa. Department of Geodesy and Geomatics Engineering University of New Brunswick.
North, D. (1990). *Institutions institutional change and economic performance.* University Press.
Nuhu, S. (2018). Peri-urban land governance in developing Countries: Understanding the role, interaction and power relation among actors in Tanzania. *Urban Forum, 30*, 1–16. https://doi.org/10.1007/s12132-018-9339-2
Owens, K. (2014). *Negotiating the city: Urban development in Tanzania.* University of Michigan.
Palmer, D., Fricska, S., & Wehrmann, B. (2009). Towards Improved Land Governance: Un-Habitat. *Land Tenure Working Paper* 11
Payne, G., Piaskowy, A., & Kuritz, L. (2014). Land Tenure in Urban Environments. Ltpr Portal: http://usaidlandtenure.net
Platteau, J. P. (1996). The evolutionary theory of land rights as applied to sub-saharan Africa: A critical assessment. *Development and Change, 27*(1), 29–86.
Un-Habitat. (2010). In *Urban land markets: Economic concepts and tools for engaging in Africa, united nations human settlements programme.* Nairobi: Un-Habitat.
Woltjer, J. (2014). A global review on peri-urban development and planning. *Jurnal Perencanaan Wilayah Dan Kota, 25*(1), 1–16.

Tinashe Bobo is both a Town Planner and a Monitoring and Evaluation Specialist. He holds a MSc and BSc in Rural and Urban Planning (UZ) and a BSScs in Monitoring and Evaluation (LSU). His research interests are in land-use planning, urban policy and management, urban informalities, housing and project monitoring and evaluation.

Audrey Kwangwama is a Lecturer at the University of Zimbabwe. Her research interests are in valuation and property development, planning and infrastructure financing. Mrs. Kwangwama is currently enrolled with the University of Zimbabwe for a Doctor of Philosophy in Planning and Real Estate.

Chapter 11
'Land Barons' as the Elephant in the Room: Planning and the Management of Urban Space in Chitungwiza and Harare

Innocent Chirisa, Abraham Rajab Matamanda, Emma Maphosa, and Roselyn Ncube

Abstract This chapter attempts to provide a nuanced discourse on the ethical implications of the clash between the so-called 'land barons', planning and the management of urban space using the cases of developments in Chitungwiza and Harare. Missing in general urban development discourse is scholarship on the interface between these three. Nevertheless, the environment of planning practice is characterised by chaos and ambivalence, especially the debate and contest among planners and politicians. The notion of 'land barons' in Chitungwiza and Harare is political. Land barons have seen the conversion of green and open spaces, or sites of delayed development (school, health, etc.) into largely low-income plots given the huge housing backlogs in these areas. Such plots become land use violations being for political rationality by political 'faces' wrestling heavy against the technical rationality championed by the planners. In the process of violations, planning has usually been rendered weak with eventual environmental consequences registered—environmental degradation, water and sanitation challenges, flooding and increasing diseases outbreaks. The quintessence of this chapter is documentation and discussion of such, being evidence, from primary and secondary data sources that have been reviewed systematically towards a policy direction that seeks to have ethics upheld and respected. If Zimbabwe is to be truly a middle-income country by 2030, this cannot be ignored.

I. Chirisa (✉)
Department of Demography Settlement and Development, University of Zimbabwe, PO Box MP167, Mt Pleasant, Harare, Zimbabwe

A. R. Matamanda
Department of Geography, PO Box/Posbus 339, Bloemfontein 9300, Republic of South Africa

E. Maphosa
Department of Architecture and Real Estate, University of Zimbabwe, PO Box MP167, Mt Pleasant, Harare, Zimbabwe

R. Ncube
Women's University in Africa, Harare, Zimbabwe

I. Chirisa and A. Chigudu (eds.), *Resilience and Sustainability in Urban Africa*, Advances in 21st Century Human Settlements,
https://doi.org/10.1007/978-981-16-3288-4_11

Introduction

This chapter seeks to map the contexts and breadth of the practices in land baronage and the associated impacts on urban planning for sustainable land development and planning ethics. This is developed in line with the Zimbabwe Agenda 2030 that seeks to transform the nation into a middle-income country by 2030 and the Africa Agenda 2063 from that most of the Zimbabwe Vision 2030 objectives are derived (African Capacity Building Foundation (ACBF), 2016; Government of Zimbabwe, 2018). The achievement of this goal hinges on land—since it has an impact on social wellbeing and the economy of cities. Thus, functional land markets and effective management of land is central to building functional cities and therefore enhancing the welfare of cities and citizens.

Conceptualising 'Land Baron'

The term "baron" was coined from the American context in that, originally, individuals from the elite and king's family were entrusted with the management of land territories within the kings' territory. Barons, later on, evolved to include not only individual from the royal family but any other person that the king would choose to entrust land. These land barons were entrusted with control of vast expanses of land and in return signed a memorandum of understanding-the Magma Carta—in that the king pledged loyalty to the barons and vice versa (Wadleu, 2008). The historical origin of land barons in America suits the definition of barons as the institutions managing land illegally on behalf of the state. The term has no single definition, but a variety of them and in some context, these definitions are in conflict. One definition is that land barons are land authorities or landlords- who own, lease or sell a large parcel of land (Editors of the American Heritage Dictionaries, 2016). The term is synonymous with real estate barons, landlords, land tycoons, land speculators or land touts depending on the context.

Though land barons are commonly conceptualised in terms of legal acquisition and subdivision of land, this study adopts a wider conceptualisation of land barons as anyone acquiring or engaged in any act related to land management where they accrue private gains at the expanse of public interest in the management of land using a specific advantage (political power relations, professional advantages or economic advantage). These activities can be in land acquisitions, land transactions, conversion or change of use of land and land valuation, skipping legal land delivery process, transactions, registration, development process as laid in the legal requirements, among other endeavours - these include those who occupy land illegally, illegal extortions and double selling) - as appropriation of public goods for private benefit. In the context of this study, land barons refer to any individual, companies engaged in an illegal practice to securing, subdividing, or development of land.

The context promoting the emergency of land barons is the rise in the urbanisation of cities wherein the local authorities are unable to address the land demand of residents concerning land supply coupled with unaccountable and self-interests in land-use planning. Other than this, land barons in countries, like America and Ethiopia (Africa) evolved from the traditional land-allocation and use framework where the king has the right to allocate portions to his subjects and these subjects assumed the responsibility to control use of such portions and collect tribute for land-use (Ambaye, 2015).

Theoretical Framework

The land management paradigm explains land administration, wherein it is not restricted to professional urban planners but to other stakeholders (the government and the private sector since land is an economic, political and social resource) (Enermark, 2005; Deininger et al., 2010). Land administration is the process of managing land including but not limited to land conveyance (registration), land transactions, land valuation and property rates collections, land development, land-use planning (site plans), determining land rights (role of courts) land-use regulations and regulating and monitoring use of land (United Nations, 1996) to achieve desired land use objectives as expressed in local development plans or a country's national development policies (Larsson, 2010). Among these goals include efficient land markets and effectively designed and functional urban spaces not only for economic welfare but also for social and environmental integrity (United Nations, 1996; Wondolleck, 1998; Deininger et al., 2010; Williamson et al., 2010). It also depicts how decisions or policies regarding land management are made by professionals and non-professionals.

Land administration is a responsibility of specific institutions where institutions refer to both actors (government, private sector - lawyers, attorney, private surveyors, urban planners, architects, real estate developers, construction companies, property valuers) and policies (UN-Habitat, 2016). These institutions have to satisfy diversified and conflicting needs of different stakeholders as far as land management services are concerned (land use, land location, land value) (Czesak et al., 2021). The assumption is that land management decisions are made rationally following a detailed analysis of the components. It is the responsibility of urban planners to balance these conflicting interests. Though these institutions play equally important roles in land administration, their power relations and relationship between these actors is also an important determinant of the effectiveness of land management.

The other component necessary for effective land management is information including cadastre systems, land demand, and land value information. The availability of land supply and demand information is one of the entry points to that land barons can take advantage of distorting information, giving false information during planning permission stages, subdivisions, distorting records or taking advantage of their information advantage to replicate the land markets (Dodson et al., 2010;

Deininger et al., 2012). However, this model was practised in developed countries and the conditions in other countries differ. In addition, the model is more suitable for management of public land (Wondolleck, 1998). The other challenge of the model is that as land management is a complex activity and as contexts, environments, legal instruments, histories, legal settings, nations globalise and demands change, so will the elements of the model and, as such, in other contexts, the model has already been modified to enhance applicability (Williamson, 2001; Williamson et al. 2010). As such, the traditional land administration paradigm is constantly adjusted, hence the development of the 'modern' land management paradigm incorporating technology and participatory approaches (Alemie et al. 2015) (Fig. 11.1).

The urban land management paradigm is necessary for understanding the study because it frames out the role of the institutions (land barons, planners, and politicians), their values in effective land management and the respective functions engaged in land administration). Institutions are the focus of the study—since barons are a form of institution that is having an important bearing on the development and management of urban land in Zimbabwe.

Urban planning is one of the tools of land administration. In the context of this study, urban planners are public land (common resource) managers who have the responsibilities to preserve the land as a common heritage for the fulfilment of community values and interests (Rouseau definition of property) (Davy, 2012). Thus,

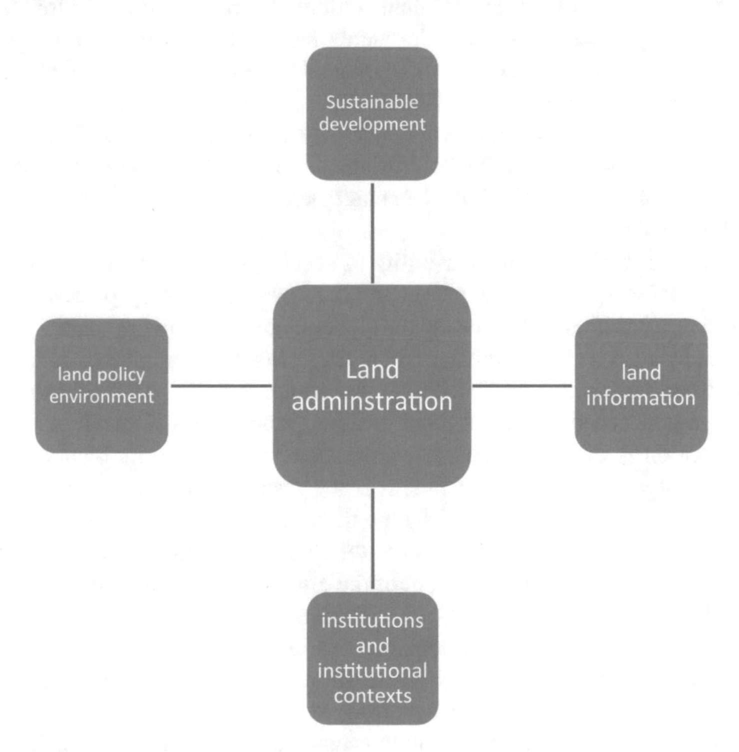

Fig. 11.1 Adopted from Deininger et al. (2010; xviii)

planners and the associated actors (elected officers) employ the public interest as a goal and criteria for approving any land-use policies and decisions (Alexander, 1992). The related land-use planning approach aligned to the public interest view is the rational comprehensive planning approach where the technical planner is assumed to be the resource person for planning and applies technical rationality in planning free from any political interferences in the decision-making process (Faludi, 1973). Understanding the public interest and the rational comprehensive planning theory is critical in realising the interests of public and private actors in land management and the rationality guiding land policy. However, the idea of a single public interest has long been disputed by scholars of the post-modernist school acknowledging a variety of interests in land-use planning that has to be acknowledged through advocacy, collaborative and communicative planning; otherwise the planning approach suffers from elite capture where planning satisfies interest of the upper class (Allmendinger, 2001; Healey, 2010; Muchadenyika, 2015). Linking politicians, planners, and other stakeholders brings a greater appreciation of the power dynamic in the planning process that shapes the practice of land administration.

Concerning urban planning ethics, every nation has got its own professional code of conduct defining the rules to be followed by members when delivering land. These codes include the iRoyal Town Planning Institute (RTPI) code, South African Council for Planners (SACPLAN) (South Africa) and the Zimbabwe Institute of Regional & Urban Planners (ZIRUP) (Zimbabwe). These codes of ethics define the acceptable moral behaviour of individuals in fulfilling their professional duties. Some of the principles of the RTPI code include competence, honesty and integrity, independent professional judgement, due care and diligence, equality and respect, and professional behaviour (RTPI, ZIRUP). However not only do the professional organisational codes affect the behaviours of actors in land-use planning, but also the individual moral and institutional values (Campell, & Marshall, 2000).

The political theory defines corruption as the abuse of political power for private gain (DasGupta et al., 2019). On the other hand, economic theory explains that rational persons are compelled into corrupt activities by economic incentives. These include payment to get a favourable interpretation of the law or discretionary treatment by inside professionals, such as policy regulators or policy journalists who are in power. This emanates from inconsistent laws or policies. In the field of new planning, corruption could arise from a deviation from fixed procedure on land use planning that is based on certainty. It involves providing monetary or other incentives to officials, such as politicians and councillors. In some cases, politicians have to pay administrative officials to maintain themselves in land administration deals (Rose-Ackerman, 1997).

There are economic consequences of corruption in land management. They include loss of efficiency and credibility in public institutions and reduction in effectiveness. Despite the costs, the public would also be benefiting. Therefore, both individuals and corrupt decisions should be weighed to evaluate costs and incentives available to those in influential positions (1991). From the perspective of economic theory, corruption involves bribery and extortion. Such economic incentives are assumed to occur between one person inside on organisation and another from outside. The insider provides information to the outsider at an economic cost. The outsider is also prepared to pay for such services (Rose-Ackerman, 1997). However, some economic arguments for corruption posit that the engagement by public officials in corrupt activities emanates from poor pay and low working conditions (Graaf, 2007).

Other than the economic opportunities, the status of agents is also an influential factor in determining engagement in corrupt activities - by shaping the motivations. These include consideration of one's status (e.g. councillors or planners) in an organisation (Andvig, 1991). Other reasons for corruption include job security issues (related to lower rank official), exposure to peer pressure (political pressure on planner from politicians or voters' pressure on councillors) and power to manipulate policies (plans, land use policies) (*ibid.*). This is related to the psychological theory of corruption where the probability of engagement in land corrupt activities varies between nations based on factors like risk-taking behaviour, the gifts- giving and taking culture, particularly in Africa. There is the moral value theory (Rusch, 2016). These can be norms of societies where; when corruption is common in society. It motivates additional corrupt practices (Gaarf, 2007). Thus, the causes of corruption can be psychological, organisational, culturally based or economically motivated.

Thus, land baronage is a form of land corruption that occur in different forms among that include theft, bribery, abuse of functions and nepotism at policy-making stages (Zuniga, 2017).

Literature Review

In Myanmar- South East Asia, large tracts of land were displaced from farmers to military personnel (Hirsch, & Scurrah, 2015). Similarly, the military is emerging as a strong land baron in Pakistan using political powers not only to acquire land and properties illegally but also protect illegal land claims of military members through intimidating local authorities in case of need to evict illegal occupiers of land (Malik, 2010). The emergence of military land barons in Pakistan is an embedded cultural practice of allocating land to the military as appreciation for service. The military have emerged as strong land barons in Pakistan through political abuse and the powers of compulsory acquisition acquired through Land Acquisition Act (Siddiqa, 2007). Other than the military, in Vietnam, business capitalists use strong power relations with local planning officials to circumvent the normal land development procedures. Local planning officials like in Australia, abuse discretionary powers to make decisions over land-use planning permits approval and conversion of land use using the rational comprehensive land-use planning approach. They manipulate decisions in land management for private gains and monetary incentives from clients (Dodson et al., 2006).

Besides abuse of political and economic powers by the elite to acquire land illegally, poor enforcement of land use regulation is also to blame. For example, in Mekong region, though land-use policy imposes limitations of the land transactions that can be allowed at subnational levels, local officials process land transactions by exceeding their powers, creating an overlap of responsibilities in land-allocation at regional and local levels. The overall effect has been the creation of opportunities for illegal transactions of land in parallel land authorities (Hirsch, & Scurrah, 2015). This is typical of Hong Kong's New Territories where, despite the existence of a formal authority responsible for managing land transactions, land-use and change of use, persists in the illegal sector (Nissim, 2012; Wang et al. 2016).

As far as the impact on land development and land-use are concerned, rather than adhering to the stipulations for developing acquired land within 18 months, the land acquired by the military in Myanmar remains undeveloped for longer periods, thus allowing speculation, rather than development (Hirsch, & Scurrah, 2015). In addition, commercial developments in cantonment areas carried out are often not properly zoned creating a scenario of conflicting land-uses in Pakistan (Siddiqa, 2007). Whilst in Myanmar, the conversion by the military is mainly for commercial development, in Hongkong's New Territories, agricultural land is illegally converted to residential property (Nissim, 2012). Consequently, all the scenarios result in a conflict of land-uses. Other than creating disharmony in land use, the other impact of the activities of military barons in Pakistan and Lahore is loss of potential revenues where rents paid for the use of land illegally are converted for commercial use are collected by the military corps (Siddiqa, 2007). Furthermore, land acquired by the militia in Lahore is often sold at prices below the fair market value- disrupting the vibrancy of formal land markets (*ibid.*).

Land baronage in Ghana is occurring in the context of growing demand for peri-urban land for commercial agriculture and residential development in towns like Kumasi and other peri-urban areas of Northern and Southern Ghana (Ubink, & Amanor, 2008; Arko-Adjei, 2011). This is occurring amidst the weaknesses of the communal land administration system that is failing to accommodate emerging urban land use needs (Yeboah et al., 2017). In addition, there is an unclear institutional framework for the responsibility for land-allocation and management between the state, chief and residents in a customary land tenure context (Ubink, 2008). This has been taken advantage of by the chiefs and planning officials. Though the chiefs are not a recognised institution bound by planning ethics, this case represents conflict of interest in the role of chiefs in the administration of peri-urban areas. Similar to the case of Ghana, multi-land use planning policies often conflicting, creating loopholes that facilitate illegal transactions and subdivision and sale of land in Kenya and Nigeria by making enforcement of land-use regulations difficult (Ghebru, & Okumo, 2016; Zuniga, 2017).

Other than conflicting legislation, the other reason promoting the emergence of land barons in Africa is elite capture in the formulation of land policy where the upper class have the power to manipulate land policies for private acquisition and benefit. This a common reason for promoting unrestricted access to land by the elite class in countries like Ethiopia, Swaziland and Nigeria since policies are made in their favour (Rose, 2002). In addition, state ownership of all public land that in return is brought under the administration of different actors (state governors, cronies and individuals provides these trustee discretionary powers for allocation, transforming these trustees to land barons and compromising effective land management for public benefit (Zuniga, 2017). Usually, these public trustees enjoy unlimited powers to regulate the use of 'state land' like in the case of state governors in Nigeria (Whitzman et al. 2013). Enjoyment of discretionary powers is prompted by modernist planning approaches that vest the power to regulate procedures for approving land use proposals to the discretion of the governor. This perpetuates land touting where the governors manipulate these powers for private accumulation of land in Nigeria (Olong, 2012). Hence the concept of the trustee holding land on behalf of 'public' interest remains a contentious issue (Cotula, 2011).

Adding to this, professional planners are also obliged to violate the principles for approving land-use applications upon receiving monetary gifts in the name of facilitation fees or informal fees from developers for planning permission. This is a common and near formal practice in Ethiopia (Addis Ababa) and Nigeria. It has the effect of degrading the recognition of land-use planning principles in approving land use applications (Burns, & Dalrymple, 2012; Ghebu, & Okumo, 2016). Thus, the absence of the rule of law in urban land-use planning is becoming prevalent.

Local chiefs administering state land in Ghana have also been benefiting when desperate land seekers pay to access residential and commercial agricultural land in peri-urban areas. These chiefs are abusing their trustee authority to sell off land or lease to rubber farmers for economic incentives (Yeboah et al, 2017). Thus, chiefs ignore their social responsibilities and accountability to citizens and pocket the proceeds from the sale of public land. In other instances, local chiefs also pledge public land to official planning authorities in return to accessing planning services and facilities for their communities (Yeboah et al., 2017). Other than chiefs, in peri-urban Mogodishabe, Botswana, individuals land occupiers also illegally subdivide land for private benefit (Home, & Lim, 2004). Besides illegal land alienation by chiefs and individuals, in Kenya and Botswana land barons take advantage of the process of change of use of public land. Using relations with politicians, these barons exert pressure on planners to approve such change of use applications for the benefit of those aligned to the politician. In some cases, false evidence is provided by planners to justify an unbefitting change of use of public land. particularly for low value uses, like agriculture to higher value uses, such as residential or commercial uses (Home, & Lim, 2004). This land value associated with a change of use accrues to private land capitalists.

Other than political pressure for approving the change of uses, there are also issues of political pressure to incorporate some prime areas into urban boundaries in the formulation of local development plans in Nigeria resulting in fake layout in council jurisdictional areas. Even in their own rights, planners have been accepting monitory gifts from elite developers for favourable treatment of applications for land-use permits. This points to a violation of not only the principle of equal treatment of all clients, but it also represents a conflict of interest within individual officials. This reflects weaknesses of the modernist planning approach where the planners are assumed to be an interest-free and neutral professional that can only work for public interests. In Kenya, the land expropriation field is one aspect in that land barons are prevalent. Government officials and capitalists abuse the law of land expropriation to expropriate public land for themselves or capitalists (Ambaye, 2015). Likewise, in Botswana, politicians employ compulsory acquisition to acquire public land for private use (Home, & Lim, 2004).

In Ghana, the impact of illegal subdivisions of land for commercial agriculture has been a violation of the minimum legally stipulated 10 hectares size of agricultural lots. This results in overworking and environmental degradation in the peri-urban (Yeboah et al., 2017). Other than the reduction in lot sizes, residential land-use has been allocated on areas intended for agriculture. This results in the misuse of land with a compromise on the quality of residential and agriculture land in Ghana. The other social implications of illegal alienation of land are the vulnerability of tenure rights of citizens securing land from barons (chiefs) since these rights are not properly registered. Since subdivisions are done privately by chiefs, revenue accrued from land sales benefits the chiefs; thus, loss of value associated with land conversion from agricultural to higher value land-uses in cities like Kumasi (Ubink, 2012). In addition, clientelistic access to public land acts as a deterrent to formal real estate market investment not only in Ghana, but in Ethiopia as well. This presents the economic costs. The other impact of land barons includes increased land conflicts, disputes and urban sprawl (Ubink, & Ammanor, 2008; Deininger et al. 2010).

Amidst the challenges posed by land barons, authorities in different countries have taken initiatives to reduce chaos bring sanity in land-use planning. Among these include the Institution of a National (urban) Land Policy in 2009 and the Land Management Programme in the early 2000s in Ghana. The Land Management Project in Ghana was meant to improve the institutional capacity by developing a civil coalition responsible for land management and improving the registry of all land under chiefs/stools land boundaries (Deininger et al., 2010). However, the initiative is prone to challenges, hence land barons persist. The national land policy was meant to streamline land-use legislation. In Ethiopia, all urban land was nationalised from ownership by a few lords to make it available and also reduce high prices (Ambaye, 2015).

Besides nationalisation, other authorities have also improved GIS techniques to improve cadastre systems and land information as a way of reducing illegal incorporations. However, nepotism and illegal subdivision of land in still persistent in Abuja, Nigeria (Akingbade, & Navara, 2010). Other attempts taken in Nigeria include promulgation of the Land Use Decree of 1978 nationalising all land within Nigeria, to curb the disjointed land administration institutions and reduce illegal land transactions outside the formal lands' office (Chukudi, 2018) However the bulk land transactions continued to occur outside the formal channels in the parallel land markets in Ibadan and Oyo States in Nigeria (Williams, 1992). To further address the issue of land touts, the government promulgated a bill in 2016 penalising illegal acquisition of land, illegal occupation, illegal alienation of property and professional misconduct by licensed professionals (Vanguard, 2016).

Methodology

This is a desktop study that employed secondary data sources including textbooks, parliamentary reports, government policies, land audits, land policies, the land corruption authorities' reports, newspaper reports, reports of law courts on land barons and journal articles written on the related issue of land barons. Though these secondary materials were not primarily directed at the topic of interest, they provided a valuable insight to the researcher by giving a diversity of views and experiences related to analysing the impact of the practice of land baronage in different contexts. These gathered views were grouped in terms of themes and analysed based on the emerging themes.

Housing Land Deals in Harare and Chitungwiza

The institutional framework and landscape for land administration in Zimbabwe is fragmented (Rakodi, 1996). There is no explicit urban land policy (save for the national land policy on agricultural land). For the comprehensive planning of land management but a variety of disjointed instruments speaking to the management of land in Zimbabwe. Typical of these is the Constitution of Zimbabwe, The Lands Commission Act, Local Land-use Masterplans, The Commission of Enquiry into the Sale of Peri-Urban State Land (dealing with acquisition and transfer of state property) and RTCP Act (Chapter 29:12) (on land development and land-use permit process, subdivisions). The emergence of land barons in urban areas in Zimbabwe is aligned to the Fast Track Land Reform period where barons emerged under the political leadership of the War Vets Associations (Moyo, & Chambati, 2013).

Drivers of Land Barons in Harare and Chitungwiza

In the context of privatisation, incremental housing delivery are some of the contextual factors fuelling the rise of land barons, in the housing land supply in Zimbabwe (McGregor, 2013). This policy removed the need for developers servicing land before selling and allowed developers to sell before developing offsite residential infrastructure (Muchadenyika, & Williams, 2017). However, the policy annulled the powers of local authorities in ensuring decent infrastructure was provided for and removing the accountability of developers by legalising the skipping of formal land development procedures. Thus, "land barons just pave the roads and collect people's money" (*Chronicle*, 2017). Adding to this are legislative bottlenecks, particularly where politicians have the final discretion concerning land-use decisions and the planner being below the politician. This promotes the culture of clientilism where politicians and councillors seek to satisfy the interests of substantial supporters (McGregor,

2013). Political powers are abused in land administration and development where companies can get away without complying with diligent land delivery, infrastructure development standards as, "some of the developers are government companies with connections and it is difficult and threaten (security of job) to regulate their activities"(McGregor, 2013: 12).

Payment and working conditions of the planners are some the issues making planners susceptible to land barons and as the latter use financial powers to persuade planners to help them provide consultation services for the development of areas that are environmentally sensitive. Though urban planners and politicians are aware of the need for preserving ethical values in the planning of urban space within cities, there are physiological challenges that make them vulnerable to corruption by wealthy land barons (McGregor, 2013), since,'some of the councillors are very hungry councillors who can hardly buy a suit, own no home, don't have the capacity to develop land" that makes it difficult to uphold ethics when some opportunities for corruption by developers or barons persist them. On the other hand, it is difficult for planners to advocate for technically rational land development for fear of harassment or loss of employment since planning officers are under the councillors (politicians) (Muchadenyika, & Williams, 2017). This shows a compromise on the ability of planners to solve the land baron issues without political support, hence perpetuation of activities of barons due to conflict of interests between planners and politicians.

Cases

Land barons' activities in the case of Manyame, Nyatsime, and St- Mary in Chitungwiza involve the illegal sale of public land on undesignated areas. In the case, Harare and Chitungwiza, the most dominant form of barons is the politically alienated individual claiming stewardship to the land. A typical example is the Chitungwiza Nyatsime Housing Cooperative (Chitungwia Municipality versus Nyatsime Beneficiaries Trust and Mavhuto Matambo and Alice Matambo, 2015). The baron allocated housing on land reserved for a cemetery.

Besides Nyatsime Housing Cooperative, another case is that of United We Stand Cooperative (Chitungwiza Municipality versus United We Stand Cooperative Fredrick Mabamba, 2013). The formal procedures are for the council to allocate land to the cooperatives. However, the experience was vice versa. The housing cooperative took the responsibilities of Chitungwiza Municipality including the subdivision of municipal land, issuing land development permits land-allocation, approval of building plans and revenue collection. These are typical of other numerous encounters of land barons in Chitungwiza and Harare. The intensity of illegal subdivisions and management of land in Chitungwiza is a reality as approximately 4000 housing lots in Chitungwiza developed illegally by barons (Parliamentary Portfolio Committee on Rural and Urban Development, 2016).

Outside the incidences of housing cooperatives within the urban confines of Chitungwiza, illegal land sales and transactions are also prevalent in peri-urban areas as demand for land increases within Greater Harare. Public land trustees including chiefs and individuals in Domboshava, Goromonzi, Seke, and Dema are therefore selling off the land and the use rights illegally a practice commonly known as Operation, "*Garawadya council isati yauya*" ("Eat before the council arrives") (Hungwe, 2014).

Political manipulation of land-use planning in Harare is also evident in the incorporation of agricultural areas, such as Hopley Farm in Harare South, Caledonia, and Hatcliffe into city boundaries. Part of Harare South, Hopley farm was co-opted into Harare City for election reasons (Mutondoro, 2018). Other than the incorporation of Hopley farm, other instances of land baronage include the case of transfer of Goromonzi to Harare City (increase in the value of farm properties for private benefits), and Charlotte Brooke and Crowhill under Goromonzi, to City of Harare. (Though Goromonzi has been retransferred back to Goromonzi Rural District Council under statutory instrument 141 of 2018).

Impacts of Land Barons

As far as the impact of land barons is concerned, "people just settled where they wanted "(PPC, 2016; 4). Not only were these chiefs and individual active, but also councillors from the Chitungwiza Council (*Sunday Mail*, May 7, 2017). This contradicts the Africa 2063 Agenda's objective to improve the quality of life by creating liveable settlements through effective land-use planning (ACBF, 2016). Furthermore, the scenario indicates abuse of office powers by councillors and planners. In most cases, these land barons are aligned to influential politicians (youth militia) or they are wealthy individuals (McGregor, 2013). In the case of politically aligned members, these have benefited from residential use of wetlands sites where the responsible authorities are unwilling to act due to political connections. Numerous cases are reported among that include the occupation of Chitubu vlei in Glen Nora, Budiriro Wetlands (*Newsday*, September 26, 2013), conversion of Hillside park (open space and wetlands) where part of Mukuvisi River drains through into a wedding venue (*Daily News*, September 24, 2018).

Though the conversion received favourable approval by responsible councillors as a permitted use under the operative masterplans, residents were against the development. The wealthy elites are able to secure preferential treatment and political support in undermining the legal procedures in land-use planning. A typical example is the construction of the Long Cheng Plaza in Belvedere, Harare by a Chinese investor where despite the provisions to preserve the site as an ecologically valuable site, the Environmental Management Agency (EMA) body has been silenced due to political pressure and development has continued (Global Press Journal, 2018). These reflect how planners are pressurised to change land-use from public to development in the name of public interest but for the benefit of politicians or business personnel (Williams and Muchadenyika, 2016). Though such change of use is permitted under the operative RTCP Act Chapter 29:12, evidence given for the change of use is not often for the public interest as stipulated by the Act but for private gain. The conversion of wetland ecosystems for development contravenes the provisions of both the Africa Agenda 2063 and the Zimbabwe Vision 2030. Both blueprints recognise the need for conservation of inland water resources for water security and environmental sustainability (ACBF, 2016; Government of Zimbabwe, 2018.

The impact of conversion of green spaces and wetlands has been the depletion of groundwater resources and increase in flash flooding on real estate property to natural floods. The risks to flooding and other environmental hazards contravene Aspiration 1 of Africa Agenda, 2063 that is environmental security. This further compromises on quality of life (ACBF, 2016). Settlement on wetlands affects ecological integrity within Greater Harare. Not only is the effect felt by households, but the Chitungwiza municipality is feeling the heat, deprived of millions of revenue, from land sales and periodic levies collected from illegal land beneficiaries. A typical example is the case of Chitungwiza Municipality (*Chronicle*, May 7, 2017). Loss of potential revenue compromises the economic standing and capacity of such local authorities (that is a contradiction to the provisions for economic resilience and competitiveness) under both the Zimbabwe 2030 Vision and the Africa Agenda 2063.

Equity in land-allocation and the issues of land supply increase is attributed to land reform, hence the major beneficiaries are ZANU PF affiliated members. This undermines the equal treatment ethics of urban planning where land has to be available to all members of society. This contradicts principles of good governance, equitable access to resources and the rule of law as in Africa Agenda 2063, Aspiration 3, where the public should acknowledge the planning authorities as professional, accountable and impartial in their discharge of responsibilities (ACBF, 2016). Similarly, it contradicts the values for observation of rule of law as in the Zimbabwe 2030 Vision (Government of Zimbabwe, 2018).

Discussion

Study findings dispute the idea of the public interest in land-use planning. Across most of the case studies of Pakistan, Mynamar, Ghana and Greater Harare, the elites are the major land barons motivated by economic incentives and the political powers setting them above the rule of law. The public interest claim is used as a legitimate strategy through that either the rich or those politically aligned to powerful politicians who secure public land for private benefit. Hence there is a need for the role of the courts to clarify the meaning of the term with regard to urban planning since it has been largely abused.

Findings from the study substantiate the value of the land-use model in analysing the loopholes in spatial management for sustainable space planning. Effective policies have to balance the interests of different stakeholder in land-use planning. However, in most of the analysed in this study, weak land management policies are a major cause for illegal land transactions and appropriation by land barons. The most typical policies taken advantage of are the public land expropriation and the change of use policies. This is typical of Ethiopia, Nigeria and also Harare and Chitungwiza in Zimbabwe. Likewise, in peri-urban lands in Ghana and Harare, land barons are taking advantage of either absence of land-use policies or conflicting urban and rural land use policies to amass and redistribute land illegally. As such, weak land-use policy is a potential cause for perpetrating the activities of land barons across African cities.

Concerning impact, land barons have negative effects on the integrity of land-use, the welfare of citizens and that of local authorities across all the case studies reviewed. This contradicts with the quest for sustainable land-use under the land management paradigm where land use planning is employed to improve the quality of life and also develop property markets in cities. As such, the ability or inability to manage land-use is central to progress or poverty of nations.

Conclusion, Policy Options and Recommendations

The study sought to explore the ethical implications of the clash between land barons, planning and the management of urban space in Chitungwiza and Harare; not only for achieving Zimbabwe's 'Middle Income Country by 2030,' but also the development of sustainable land-use planning. Findings reveal that the notion of 'land barons' in Chitungwiza and Harare is political. Land barons have seen the conversion of green and open spaces, or sites of delayed development (school, health.) into largely low-income plots given the huge housing backlogs in these areas. Such areas become land-use violations against the technical rationality championed by urban planners. Other than haphazard development, the other effect has been environmental degradation, water and sanitation challenges, flooding and increasing diseases outbreaks—when wetlands reserved for ecological purposes are changed into residential use. This presents a persistent cost to the low-income earners as they suffer water, sanitation and flooding challenges but also insecure land tenure as in most cases their claims to land remain unregistered at law. Other than a loss to citizens, the local authorities formally responsible for managing development and use of land in such areas also suffer the loss of potential revenue from both land sales and land value increments from a change of use at the expense of the baron. The effect has been an erosion of not only the internal integrity of the planning profession, but also citizens who have long questioned the value of the presence of urban planners.

In line with the major findings and causes of land baronage in Zimbabwe and the experiences across the globe, the following recommendations are proposed to reduce the occurrence of land barons in Harare and Chitungwiza:

The institutionalisation of a national urban land policy can help in amalgamating the various policies and actors and coordinating activities of the urban land markets in Harare and Chitungwiza and their peri-urban areas, the lack of which led to the prevalence of barons. A national urban land policy would also compliment the activities of the Lands Commission and the Commission of Enquiry into the Sale of Peri-urban State Land. In addition, development of democratic urban land-use planning contexts would help reduce the high discretion of councillors and planners in falsifying public interest for private gain, particularly in the change of use of public land and open spaces within the metropolitan. This is because the area of public interest in land-use conversions and acquisitions remains controversial in the allocation of public land to individuals aliened to powerful politicians in the areas of study.

In line with the public interest, there is also need for reclassification of the definition of public interest as implied in the laws for a change of use in the RTCP Act Chapter 29:12, since the legal definition has been monopolised to serve interests of the elites in land acquisition and use. The other policy proposal is the institution of a solid urban planners' act to regulate the activities of practitioners despite the existence of the ZIRUP professional body of planners. This is because planners are also abusing the planning powers in return for benefits from land barons or they are also engaged as land barons in their own capacities- that reflects a conflict of interest and violation of ethical conduct. Thus, unless the unethical planners and politicians are tamed, the problem is likely to persist.

References

African Capacity Building Foundation (Acbf). (2016). *African Union Agenda 2063*. S.L.: Acbf.

Akingbade, A., & Navara, D. G. Y. (2010). the impact of Abuja Gis on corrupt practices in land administration in the federal capital territory of Nigeria. *Ict and Development, 9,* 1–16.

Alemie, B. K., Bennett, R. M., & Zevenbergen, J. (2015). A socio-spatial methodology for evaluating urban land governance: The case of informal settlements. *Journal of spatial science, 60*(2), 289–309.

Alexander, E. R. (1992). *Approaches to Planning: Introducing Current Planning Theories, Concepts and Issues* (2nd ed.). Australia: Gordon and Breach.

Allmendinger, P. (2001). *Planning in Post Modern Times*. London: Routledge.

Ambaye, D. (2015). *Land Rights and Expropriation in Ethiopia*. New York: Springer International.

Andvig, J. (1991). The economics of corruption: A survey. *Studi Economici, 43*(1), 57–94.

Arko-Adjei, A. (2011). *Adapting land administration to the institutional framework of customary tenure*. The Case of Peri-Urban Ghana. PhD Thesis: University of Twente.

Burns, T., & Dalrymple, K. (2012). Land sector corruption in Ethiopia. In J. Plummer, (ed.), *Diagnosing Corruption in Ethiopia: Perceptions, Reality and The Way Forward for The Key Sectors*. Washington D.C: World Bank. pp. 285–326.

Campell, H., & Marshall, R. (2000). Moral obligations, planning and the public interest. *Environment and Planning B, 27,* 297–312.

Chitungwiza Municipality Versus Nyatsime Beneficiaries Trust and Mavhuto Matambo And Alice Matambo (2015).

Chitungwiza Municipality Versus United We Stand Cooperative Fredrick Mabamba (2013).

Chronicle. (2017). *Land Barons Menace Worrying: Kasukuwere,* S.L.: S.N.

Chukudi, V. (2018). *Problems and Prospects of Urban and Regional Planning in Nigeria: Port Harcourt Metropis since 1914*. New York: Page Publishing.

Cotula, L. (2011). *Land deals in Africa: What is in the contracts?* London: IIED.

Czesak, B., Różycka-Czas, R., Salata, T., Dixon-Gough, R., & Hernik, J. (2021). Determining the Intangible: Detecting Land Abandonment at Local Scale. *Remote Sensing, 13*(6), 1166.

Daily News. (2018). *Hillside Park Converted to Wedding Venue,* S.L.: S.N.

DasGupta, R., Hashimoto, S., Okuro, T., & Basu, M. (2019). Scenario-based land change modelling in the Indian Sundarban delta: An exploratory analysis of plausible alternative regional futures. *Sustainability Science, 14*(1), 221–240.

Davy, B. (2012). *Land Policy: Planning and the Spatial Consequences of Property*. Burlington: Ashgate.

Deininger, K., Augustinus, C., Enemark, S., & Munro- Faure, P. (2010). *Innovations in Land Rights Recognition, Administration and Governance.* Washington Dc: World Bank.

Deininger, K., Savastano, S., & Carletto, C. (2012). Land fragmentation, cropland abandonment, and land market operation in Albania. *World Development, 40*(10), 2108–2122.

Dodson, J., Ellway, C., & Coiacetto, E. (2006). Corruption in the Australian land development process: identifying a research Agenda. *Governance, 09,* 1–25.

Dodson, J. (2010). In the wrong place at the wrong time? Assessing some planning, transport and housing market limits to urban consolidation policies. *Urban Policy and Research, 28*(4), 487–504.

Editors of the American Heritage Dictionaries. (2016). *The American Heritage Dictionary of the English Language.* 5th Ed. S.L. Houghton Mifflin Harcourt Publishing.

Enermark, S. (2005). *Land Management for Institutional Development* (pp. 1–14). S.N: Melbourne.

Faludi, A. (1973). *Planning Theory*. Oxford: Pergamon Press.

Ghebru, H., & Okumo, A. (2016). *Land Administration Service Delivery and Its Challenges in Nigeria: A Case Study of Eight States.* S.L.: International Food Policy Research Institute.

Global Press Journal. (2018). *Zimbabwe Wetlands Threatened as Illegal Development Goes Unpunished,* S.L.: S.N.

Government of Zimbabwe. (2018). *Government of Zimbabwe 2018 Transitional Stabilisation Programme: Reforms Agenda October 2018—December 2020.* Harare: S.N.

Graaf, G. (2007). Causes of corruption: Towards a contextual theory of corruption. *Paq Spring,* 39–86.

Healey, P. (2010). *Making Better Places: The Planning Project in the Twenty-First Century*. Hampshire: Palgrave Macmillan.

Hirsch, P., & Scurrah, N. (2015). *Land Grabbing, Conflict and Agrarian- Environmental Transformations: Perspectives from Eastern and Southern Asia.* S.L., International Academic Conference Paper No. 48.

Home, H., & Lim, H. (2004). *Demystifying the Mystery of Capital: Land Tenure and Poverty in Africa and the Carribean.* Orogen: Cavendish Publishing.

Larsson, G. (2010). *Land management as public policy*. University Press of America.

Hungwe, E. (2014). *Land transactions and rural development policy in the Domboshava Peri—Urban Communal Area.* Zimbabwe, S. L.: Stellenbosch University.

Malik, A. (2010). *Political Survival in Pakistan: Beyond Ideology*. London: Routledge.

McGregor, J. (2013). Surveillance and the City: Patronage, Politics and Urban Control in Zimbabwe. *Journal of Southern African Studies, 39*(4), 783–805.

Moyo, S., & Chambati, W. (2013). *Land and Agrarian reform in Zimbabwe: beyond white settler capitalism.* Senegal: Codesria.

Muchadenyika, D. (2015). Land for housing: A political resource—reflections from Zimbabwe's Urban Areas. *Journal of Southern African Studies, 41*(6), 1219–1238.

Muchadenyika, D., & Williams, J. (2017). Politics and the Practice of Planning: The Case of Zimbabwe's Cities. *Cities, 63,* 33–40.

Mutondoro, F. (2018). *Housing Cooperatives, Political Symbols and Corruption in Urban Land Governance: The Case of Zimbabwe.* S.L.: Orangetree.

Newsday. (2013). *Police Descend on Urban Land Invaders,* S.L.: S.N.

Nissim, R. (2012). *Land administration and practice in Hong Kong* (3rd ed.). Aberdeen: Hongkong University Press.

Olong, A. (2012). *Land Law in Nigeria.* Lagos: Malthouse Press.

Parliamentary Portfolio Committee on Rural and Urban Development (Ppcrud). (2016). *First Report of the Parliamentary Portfolio Committee on Rural and Urban Development 2016 On Service Delivery in Local Authorities,* Harare: Government of Zimbabwe.
Rakodi, C. (1996). Urban land policy in Zimbabwe. *Environment and Planning A, 28*(9), 1553–1574.
Rose, L. (2002). African Elites Land Control Manoeuvres. *Etudes Rurales* (163–164), 187–213.
Rose-Ackerman, S. (1987). Ideals versus dollars: Donors, charity managers, and government grants. *Journal of Political Economy, 95*(4), 810–823.
Rose- Ackermn, S. (1997). The political economy of corruption. *Corruption and Global Economy, 31*(1997), 31–60.
Rose-Ackerman, S. (Ed.). (2007). *International handbook on the economics of corruption.* Edward Elgar Publishing.
Rusch, J. (2016). *The Social Psychology of Corruption.* S.L., Oced, pp. 1–22.
Siddiqa, A. (2007). *Military Inc.: Inside Pakistan's military economy.* New York: Oxford University Press.
Sunday Mail. (2017). *Chitungwiza Councillors in Us $7 m Land Deals,* S.L.: S.N.
United Nations. (1996). World Urbanization Prospects: The 1996 Revision. United Nations, New York, NY.
Ubink, J., & Amanor, K. (2008). *Contesting Land and Customs in Ghana: State, Chief and the Citizen.* S.L.: Leiden University Press.
Ubink, J. (2008). *Traditional authorities in Africa. resurgence in an era of democratisation.* Leiden: Leiden University Press.
UN-Habitat. (2016). World Cities Report, Nairobi: UNHABITAT.
Vanguard. (2016). *Law against Touts Most Welcome.* Available Online: https://www.vanguardgr.com. Accessed 22 february 2019.
Wadleu, J. H. (2008). *Law, Investigation and the Collective Bargaining Agreements.* Bloomington: Authorhouse.
Wang, A., Chan, E., Yeung, S., & Han, J. (2016). Urban fringe land use transitions in Hong Kong: From new towns to new development areas. *Procedia, Engineering, 198*(2017), 707–719.
Whitzman, C., Legacy, C., & Andrew, C. (Eds.). (2013). *Building inclusive cities: Women's safety and the right to the city.* Routledge.
Williams, D. (1992). Measuring the impact of land reform policy in Nigeria. *Journal of Modern African Studies, 30*(4), 587–608.
Williamson, I., L, Ting., & Grant, D. (1999). The evolving role of land administration in support of sustainable development: A review of the United States—international federation of surveyors barthurst declaration on land administration for sustainable development. *Australian Surveyor, 44*(2), 126–135.
Williamson, I., Enemark, S., Wallace, J., & Rajabifard, A. (2010). *Land Administration for Sustainable Development*. Redlands: Esri Press Academic.
Williamson, I. P. (2001). Land administration "best practice" providing the infrastructure for land policy implementation. *Land Use Policy, 18*(4), 297–307.
Wondolleck, J. M. (1998). *Public lands conflict and resolution: Managing national forest disputes.* New York, NY: Plenum Press.
Yeboah, E., Addda, M., & Okai, M. (2017). *Women, Land and Corruption in Ghana: Findings from a Baseline Survey.* Washington Dc, World Bank, pp. 1–25.
Zuniga, N. (2017). *Land Corruption Topic Guide.* S.L.: Transparency International.

Innocent Chirisa is a Full Professor in Environmental and Regional Planning at the University of Zimbabwe. At the present time, he is the Dean of the Faculty of Social and Behaviural Sciences at the University of Zimbabwe. He is the Deputy Chairman of the Zimbabwe Ezekiel Guti University. He has served as the Deputy Dean of the Faculty and Chairman of the Department of Rural and Urban Planning at the same University. His keen interest is in urban and peri-urban dynamics. Currently focusing on environmental systems dynamics with respect to land-use, ecology, water and energy. Professor Chirisa is a seasoned scholar who also contributes in a local Sunday Mail newspaper

Abraham Rajab Matamanda is a Lecturer in the Department of Geography, University of the Free State, South Africa

Emma Maphosa is an MSc (Rural and Urban Planning) student in the Department of Architecture and Real Estate, University of Zimbabwe

Roselin Ncube is a lecturer at Women 's University in Africa. She is currently enrolled with the same University for a Doctor of Philosophy Degree (D.Phil.). She holds MSc (Social Ecology) from the University of Zimbabwe and a B.Sc. Honors in Sociology and Gender Studies from Women 's University in Africa. Roselin has co-authored a number of chapters in published books. Her lifelong ambition has always been achievement of the penultimate academic excellence in social sciences with a special niche in Gender Issues, Urban Sociology, Women Empowerment and Entrepreneurship and Urban Agriculture.

Chapter 12
Climate Change and the Resilience Cause in Masvingo City Urban Landscape, Zimbabwe

Washington M. Mbaura

Abstract Despite many global agreements made on climate change, little has been done on the provision for landscape resilience to climate change in Masvingo City. The aim of this research is to examine landscape resilience to climate change in Masvingo City to enable it to navigate change, build local capacity to respond to disturbance and prepare for uncertainty and foster a transition to more sustainable urban trajectories. A review of literature was done to gain insight on what other scholars wrote concerning the urban landscape resilience under the impact of climate change. A mixed methods approach was used whereby interviews were done to solicit data from town planners in Masvingo City Council, Department of Physical Planning and Environmental Management Agency and observations and focus group discussions were conducted to complement data from interviews. Purposive sampling technique was used to select research subjects. The findings exhibit that existing urban landscapes partially meet the principles of resilience to climate change and that there is lack of regulatory frameworks to provide for landscape resilience to climate change. Therefore, the key findings point at the need for urgent action towards the establishment of a planning legislation covering landscape resilience to climate change.

Introduction

Though the world cities only cover 2% of the earth's surface, half of the world's population is living in urban areas (Leichenko, 2011). Increasing economic, social and spatial vulnerabilities and pressures due to incorporation of urban areas into the new global economy and opening the door to external pressures necessitate building resilient urban systems. Moreover, increasing landscape vulnerabilities requires connecting planning and science of ecology, thereby enhancing landscape resilience of urban systems. Whilst the majority of previous studies concentrated

W. M. Mbaura (✉)
Department of Demography Settlement and Development, University of Zimbabwe, Mt Pleasant, PO Box MP167, Harare, Zimbabwe

I. Chirisa and A. Chigudu (eds.), *Resilience and Sustainability in Urban Africa*,
Advances in 21st Century Human Settlements,
https://doi.org/10.1007/978-981-16-3288-4_12

on considering the available information of the resilience concept, there seems to be little study done on enhancing urban landscape resilience under the impact of climate change in urban areas, particularly in developing countries where cities are growing rapidly and a high proportion of urban populations are poor or otherwise vulnerable to climate-related disruptions.

Most of the developing world cities have been designed without consideration of climate change, as it once seemed like a static factor (UN-Habitat, 2010). The ever-increasing need to build capacity for greater resilience will require cities to develop strategies for coping with the future shocks and stresses to urban infrastructure systems associated with climate change. A changing climate will also affect city planning, in that areas may become unsuitable for development due to increased risk. Considering the available information on the resilience concept, there exists a gap in recent studies, mainly in regard to how to improve and promote resilience of urban landscapes in light of climate change. Consequently, this perception of climate change by previous urban planners left most cities with little to no resilience to the changing climate and thus posing dangers to human lives and implying the need for its consideration. It is against this background that this study intends to bring out ways of enhancing urban landscape resilience under the impacts of climate change that have been experienced in Masvingo City. This chapter examines landscape resilience to climate change in Masvingo City to enable the city to navigate change, build local capacity to respond to disturbance and prepare for uncertainty and foster a transition to more sustainable urban trajectories.

Masvingo City was used as a case study and the city has various urban landscapes that include the built environment and the natural. This study covered provisions for enhancing urban landscape resilience under the impact of climate change because scholars that preceded this study tended to generalise climate change and its impacts in the continent, rather than a specific-context, like Masvingo, thereby making their studies too broad.

Background to the Study

The climate is changing the world over and the need to address the impacts on human and natural systems is becoming more urgent with increasing scientific evidence. While planning for climate change has been predominantly focused on mitigation; adaptation is coming to the forefront of climate change planning, especially within the context of urban areas. Urban regions are vulnerable and are already affected by climate change. This is due to emissions, increased exposure and risk and the deficiency of existing urban systems to cope with known and uncertain magnitude of impacts that render urban regions and their populations vulnerable to changing precipitation regimes, increased temperatures and rising sea level (Linkov et al., 2014). As such, cities are starting to address these impacts through initiatives that change people's preference for living (denser developments, reduction of power demand). There is need for the establishment of resilient cities, that can absorb changes, manage climate induced risks and recognise environmental, social and economic development in an integrated manner.

Since the 1980s, the mass media influenced national policy responses by their extensive coverage of global warming and related issues that created public concern and a call for political commitment. Climate change became a matter of scientific and policy attention, and debates surrounding climate change have focused on the challenge of mitigation (Weingart et al., 2000). The issue has mostly been on how to promote and govern a transition towards sustainability (Walker et al., 2004). Even though the normative concept of sustainable development is already well established at the global, transnational, national, regional, and local levels of politics, concrete steps towards more sustainable societies are still lagging far behind the intentions of the concept and also behind the expectations established at international summits (Keiner & Kim, 2007). In addition, the global climatic system and human society are continuously changing systems. They sometimes evolve in response to impacts emerging from the other system and sometimes they evolve autonomously (Keiner & Kim, 2007). A changing, uncertain world demands action to build the resilience of the social-ecological systems that embrace all of humanity (Folke et al., 2002).

In 1979, the First World Climate Conference by the World Metrological Organisation highlighted, among other topics, the necessity for nations across the world to predict and prevent potential man-made changes in climate that might impact the well-being of humanity (Markard et al., 2012). In the wake of this conference, the World Climate Research Program was established to determine the predictability of climate and the effects of human activities on it (Ernstson et al., 2010). This led the United Nations Environment Program (UNEP) and the World Metrological Organisation to be founded in 1988, and the Intergovernmental Panel on Climate Change (IPCC) (Weingart et al., 2000). The IPCC published its First Assessment Report in 1990 at the Second World Climate Conference. Like its predecessor, this conference highlighted the risks of climate change and led to the establishment of the United Nations Framework Convention on Climate Change (UNFCCC). Under UNFCCC's auspices, the 1997 Kyoto Protocol sought to obligate mostly industrialised countries to reduce their greenhouse gas emissions (UNFCCC, 2013). The 1980s and 1990s also witnessed the endorsement of several reports and agreements, such as IPCC's First and Second Assessment reports (in 1990 and 1996 consecutively), the Kyoto Protocol, and Agenda 2013. Thus, during these decades, climate change was initiated as a scientific field that set new directions for researchers from various fields, while several climate change-related notions developed, including the aforementioned adaptation and mitigation and vulnerability, risk, and resilience.

Tyler and Reed (2011) identified the different foci of adaptation studies, including ones that tackle the fundamental definitions related to climate change and governance and institutional policies. Their study also revealed that the steadily growing body of literature on climate change adaptation has rarely included spatial planning, urban form, or urban design. Others have also established that the majority of adaptation studies thus far underscore isolated aspects of adaptation and rarely address the complex interactions between the bio-physical, economical, and environmental agents of an urban area (Lankao, 2010). In fact, the climate change adaptation literature, generally, acknowledges this inadequacy and hence recommends better urban planning and design underscores the quality of building, infrastructure, and services and advocating for land-use planning and management that would collectively enhance the resilience of urban areas (World Bank, 2011). Notably though, while resilience has been discussed in the socio-ecological literature since the 1970s, it remains a relatively new concept in urban planning discourse, particularly vis-à-vis climate change adaptation (Davoudi et al., 2012).

Decision-makers, practitioners, and scholars are now prioritising Climate Change Adaptation (CCA) in their agendas, resulting in a rapid rise in the number of adaptation related publications in the last decade (Ernston et al., 2010). Consequently, CCA is now an important research field (UNISDR, 2012). The Intergovernmental Panel on Climate Change (IPCC) Fourth Assessment Report (4AR) has been instrumental in shaping the direction of current adaptation research (IPCC, 2007) from the biophysical to the social, economic and institutional aspects of adaptation (Davoudi, 2012). This change has prompted researchers to focus on the linkages between CCA and sustainable development. Sustainable development can reduce vulnerability to climate change, and climate change could impede the nations' abilities to achieve sustainable development pathways (IPCC, 2007). Hence, development and adaptation are acknowledged as mutually dependent strategies, and efforts to streamline climate related concerns into the development planning and decision-making processes (Linkov et al., 2014). Along with this is the change in the focus of CCA planning from impact, vulnerability, or risk assessments to adjusting the direction in development planning.

Literature Review

The Climate Resilience Framework (CRF) provides a structure for vulnerability analysis that combines systemic, behavioural, and institutional factors. This enables insights into the sources of resilience and vulnerability that are different from those generated through either human-centred approaches or those based on systems analysis alone. By disaggregating approaches to analysis and drawing attention to the interactions between the behavioural drivers of agents, institutional contexts, system dynamics, and changing patterns of exposure, it enables identification of targeted points of entry at different scales where interventions can build resilience. These relate to characteristics (such as diversification, flexibility and the ability to learn) that are fundamental to both human and systems resilience. In conjunction with shared-learning and iterative planning, this creates an adaptive mechanism for translating insights into practical courses of action over time. For the framework to be useful, however, considerable skill is required in adapting its use to different contexts. Substantial translation and interpretation are required for use with practitioners or policy actors. Finally, work is required to clarify how many of the characteristics that contribute to resilience are hazard-, culture-, or otherwise specific. While evidence suggests that the basic characteristics of systems, agents, and institutions that contribute to resilience are generalisable, such frameworks cannot be applied blindly.

The initial references to the notion of resilience in planning dates back to 1990s (Mileti, 1999), the core idea of cities' ability to respond to internal and external shocks and perturbations, such as natural disasters are not new for urban planning (Batty, 2013). Resilience in urban planning is grounded in four different areas of research: urban ecological resilience, urban hazards and disaster risk reduction, resilience of urban and regional economies, and urban governance and institutions that promote resilience (Leichenko, 2011). Yet, "urban resilience" as a new concept is mainly established in the context of environmental hazards focusing on disasters, communities, and most recently on the impacts of climate change including climate-related disasters (Chelleri et al., 2015; Davidson et al., 2016). As a result, the literature on urban resilience mainly overlaps with disaster resilience and community resilience literature (Davidson et al., 2016).

Timmerman et al. (1981) defined resilience as the measure of a system's capacity to absorb change and recover from occurrence of hazardous events. This includes both natural and human-induced disasters, such as terrorist attacks (Burby, 2009; Coaffee, 2008; Godschalk et al., 2003). Resilience in disaster research is mainly focused on engineered and social systems, including measures to prevent the damage caused by disasters and post-event strategies to cope with or reduce impacts and aid recovery (Tierney and Bruneau, 2007). To be resilient to disasters, therefore, means to be flexible in coping with such events and be able to make the best of the opportunities that might arise (Müller, 2011). Community, as the totality of social systems within a defined geographic boundary (Cutter et al., 2008), has been a central focus in the literature on disaster resilience.

In accordance with the conceptual evolution of resilience, there are three distinct approaches to conceptualising urban resilience namely static urban resilience, socio-economic urban resilience and evolutionary urban resilience (Davidson et al., 2016). Static urban resilience is the ability of an urban system or city to withstand shocks and return to its previous state and is the equivalent of engineering understanding of resilience. Socio-ecological urban resilience is the ability of an urban system or city to adjust to shocks and stresses and minimise disruption by reorganising its parts; and evolutionary urban resilience is the ability of an urban system or city to learn, adapt, and transform in response to shocks. Some researchers argue that human systems can never return to their original state due to the social learning process and social memory that they entail (Magis, 2010). Furthermore, a return to a previous state is not always ideal for an urban system considering the undesirability of vulnerabilities that essentially have led to disasters (Manyena et al., 2011).

There are, however, contradictory definitions of urban resilience in the literature. While some researchers define it as "the degree to that cities tolerate alteration before reorganising" (Alberti, 2003:5) or "the ability to withstand a wide array of shocks and stresses" (Leichenko, 2011:71), that conceptualise resilience as a return to normalcy, others consider it as the "ability to absorb, adapt, and respond to changes in urban systems" (Desouza & Flanery, 2013:22). Such conceptual flexibility allows for interdisciplinary bridging and integration by creating shared understandings without the need for precise definitions (Brand & Jax, 2007), but it can also lead to significant confusion and vagueness in practice by allowing decision-makers to interpret resilience based on their interests or areas of expertise (Lu & Stead, 2013). For instance, some consider resilience as a synonym for vulnerability, adaptation, or sustainable development, even though there are significant differences between these concepts (Elmqvist et al., 2014, Meerow et al., 2016). Such confusion can pose a barrier for operationalising resilience and measuring progress towards it while limiting opportunities for further collaborations across disciplines (Brand & Jax, 2007; Leichenko, 2011).

The terminology of resilience, exposure and vulnerability is widely used across several related fields, but with little consistency or consensus on definitions (Berkes, 2007; Gallopin, 2006; Klein et al., 2003). Low resilience systems are intrinsically vulnerable to stress and shock, so in this sense increasing resilience reduces vulnerability (Folke, 2006).Within this conceptual framework, building urban climate resilience means strengthening systems to reduce their fragility in the face of climate impacts and to minimise the risk of cascading failures, building the capacities of social agents to anticipate and develop adaptive responses, to access and maintain supportive urban systems and addressing the institutional factors that constrain effective responses to system fragility or undermine the ability of agents to take action.

Subsequently, urban resilience refers to learning, planning, forecasting, resisting, absorbing, accommodating to and recovering from unforeseen changes within cities (Jabareen, 2012). City resilience is a complex and multidisciplinary phenomenon, it requires cooperation between various actors working within governance, spatial, economic and social urban dimensions (Ibid.) It is an emerging concept within both urban planning and designing, and requires increasing efficiency, such as sustainable energy production; decreasing reliance of oil and non-renewable resource consumption and localising economic development (Arefi, 2011). Resilience of an urban system is dependent on the interactions between the urban system and the natural environment. Urban ecological resilience is rooted in landscape ecology. Ahern (2011) considers landscape ecology as an interdisciplinary field that defines landscapes and heterogeneous spatial entities with inherent disturbance regimes depending on type, frequency, and intensity of the recognised disturbances. Landscape ecology is connected to landscape planning in terms of landscape form, patterns, and change (Turner, 1990). Building resilience through landscape and planning requires the identification of disturbances that a particular landscape or city is likely to face, the frequency and intensity of these events, and how cities can build adaptive capacity to respond to these disturbances while remaining functional (Vale et al., 2005).

Urban systems rely heavily on the functioning of their infrastructure systems and it is usually these systems that fail during disasters and put many human lives at risk of climate-related events. For example, transportation systems simultaneously affect relief logistics, emergency medical treatment, accessibility, and recovery from disasters (Hsieh & Feng, 2014). Infrastructural resilience has an important economic dimension. The main infrastructures of cities in many developing and developed countries are potentially vulnerable to future climate-related events. For example, in Australia more than $226 billion of commercial, industrial, road and rail, and residential infrastructure are vulnerable to inundation and erosion hazards in face of future sea level rise (Department of Climate Change and Energy Efficiency, 2011). Resilience in this context can be achieved by lowering possibilities of failure (before the event), reducing severe negative consequences in case of failure (during the event), and enhancing a faster recovery rate (after the event) (Bruneau, 2003). When applied to cities, urban infrastructural resilience goes far beyond overcoming technical barriers. Rather, it has a very important societal dimension to consider since infrastructure failure is a primary cause of economic and human losses during disasters (Chang, 2009; Simonsen et al., 2014).

In the perspective of cities and resilience thinking, it is important to notice that these principles can be strengthened by global interaction. Ernstson et al. (2010) describe it as cross-scale interaction, that can be a key in driving changes in slow variables to push urban systems across thresholds. In addition, Bathelt et al. (2004) mention the importance of enhancing knowledge by investing in building channels of communication to selected providers located outside the local milieu.

The population of Africa's cities is growing rapidly. However, as poor people cram into towns and cities characterised by limited, weak and often under-resourced infrastructure, they are increasingly relegated to marginal, inadequately serviced, informal settlements and low-cost housing areas, leaving them vulnerable to numerous livelihoods, health and security risks. Disasters undo progress in achieving developmental goals, such as gains in education, healthcare and economic progress, and prevent vulnerable women, men and children from being able to claim their rights to basic needs, such as food, shelter, work and healthcare. They also erode individual and household resources, undermining livelihoods and the realisation of human rights, that, in turn, increases vulnerability to disasters of all magnitudes.

Urban vulnerability includes both biophysical and socio-economic dimensions. The biophysical exposure of a population is determined by the probability of a hazardous event and the proportion of the community that will be affected (Cross, 2001). The location, geography, and size of settlements, the built environment and infrastructure, the natural resources at risk, and sea level rise are important factors contributing to biophysical vulnerability. Socio-economic vulnerability is the inability of a population to cope with societal and environmental hazards and stresses (Lankao & Qin, 2011). Population growth, demographic structure, level of income, education, health, social capital, economic vulnerability, and governance are among the main socio-economic factors contributing to urban vulnerability. Understanding both the biophysical and socio-economic underpinnings of vulnerability is crucial for responding to climate change (Preston, 2013).

Risk, by definition, is the chance of occurrence of an undesirable outcome (ISO31000 2009). The most common formula describing risk adds the likelihood of a hazard to the vulnerability of a population (Preston & Stafford-Smith, 2009). Hazards, such as floods and cyclones, as the source of harm, might not necessarily pose a high risk for an area on their own. yNonetheless, it is the combination of these hazards with the vulnerability of the population, settlements, and infrastructure that can create disasters (Wamsler, 2014). The IPCC's special report on managing extreme events (IPCC, 2012) considers disaster risk as a combination of hazard, exposure, and vulnerability. Based on these risks, urban environments can respond to climate-related disasters by building resilience. Failure to adapt, or become locked in maladaptive patterns that further lead to disasters could worsen the situation. There is mounting international concern about how to address the implications of climate change for urban areas, particularly in developing countries, where cities are growing rapidly and a high proportion of urban populations are poor or otherwise vulnerable to climate-related disruptions.

Increasing economic, social and spatial vulnerabilities due to incorporation of urban areas into the new global economy and opening the door to external pressures also necessitate the building resilient urban systems. The entrepreneurial logic in property markets decreases the opportunity for public concerns, and unequal power relations and the privatization of the state make proactive measures to unexpected crisis and hazards difficult. Moreover, increasing ecological vulnerabilities requires connecting planning and science of ecology and enhancing ecological resilience of urban systems, and considering the impact of already-foreseen or unforeseen threats to ecosystems.

Urbanisation can amplify the likelihood of disasters, while the concentration of people and infrastructure increases exposure to hazards and the magnitude of their impact. Urban expansion alters land surfaces, disrupts natural processes and systems, aggravating flooding, for example, by covering ground with hard surfaces, such as roofs, roads and pavements. These do not absorb rainwater as vegetation does, thereby increasing run-off and pooling. Activities, such as road construction, pollution, wetland reclamation for residential and commercial use and resource extraction, also diminish ecosystem services, such as flood regulation and protection. In rapidly-expanding towns and cities, space is at a premium. In the absence of affordable, well-located land, poor households often have little choice but to live on marginal land—cheap or vacant precisely because it is unsafe. In many countries, urban population growth has outpaced the capacity of the authorities to maintain and expand infrastructure and provide essential services. Shortages in affordable housing have also resulted in the growth of large unplanned informal settlements across Africa.

Climate change is aggravating existing risks and creating new ones. The effects of climate change in Africa are as varied as its diverse environments and climates, but include changing temperatures, more intense storms, variable and unpredictable weather and increased occurrence of drought. UN-Habitat (2010) states that climate change presents an immediate threat for African cities, as many large cities lie in coastal areas and are, especially vulnerable to rising sea levels, saline penetration, storm surges, flooding and coastal erosion. Urban populations are also affected by changes in the availability and cost of food, water, energy and transport. Adverse environmental conditions, such as drought also increase migration to urban areas, as rural livelihoods suffer.

The concept of resilience has recently been applied to that of cities by researchers, urban planners and local governments alike. Examples of this are new urban resilience models, such as "the eco-efficient city", the "carbon–neutral city" or "the place-based city" (Arefi, 2011). Urban areas are in a constant process of internal and external change, that is, they decline or expand, developing new form and function, dealing with various difficulties, such as segregation, changing demographics and spatial patterns, economic crisis and global competition (Marcuse & van Kempen, 2000). They never fully enter a state of stability, and keep on shifting causes, appearances, scales and effects (Ibid.). In this sense, urban areas are similar to other human or natural systems. Arefi (2011) argues that cities and organisms have two key elements in common: the ability to recover from a disaster (or an illness), and the ability to absorb and adapt to change (Arefi ibid.). This shows that the concept of resilience is possible to apply within urban studies. It has also come to be regarded as a necessity to reduce the negative impacts of the above-mentioned changes, and instead increase the safety and wellbeing of cities and their residents (Jabareen, 2012). For a city or urban community that is not striving towards increased resilience is more vulnerable to risks and threats, affecting areas, such as water, food, energy, infrastructure, flows of goods and services and residential health and safety (Arefi, 2011). This has become even more urgent as of late since cities, around the world are being tested more frequently and by greater intensity from impacts such as climate change (ICLEI, 2012).

A resilient system is defined by two main features that is; its ability to absorb change and disturbance, and the persistence of systems while retaining its basic functions and structure (Walker et al., 2006); together with the ability to survive, adapt and transform itself (Ludwig et al., 1997). The attributes above define a possible choice in building a planning framework on whether to follow conservative or radical constructs of resilience (Raco & Street, 2012). The former view of resilience allows a return to the steady state that existed before the external shock threatened to bring radical and fundamental change, while in contrast the latter interpretation sees resilience as a dynamic process involving the rejection of the status quo, as there can be no return to the circumstances that actually caused the problem in the first place (Raco & Street, 2012). The latter definition, accepted here as the core of the resilience planning paradigm, can be defined with respect to three aforementioned dynamic assets of the urban systems: adaptive capacity, self-organisation and transformability, rather than characteristics connected to the steady-state condition.

Evaluating urban systems with respect to these assets enables one to determine the critical issues for resilience planning. First, it has to be dynamic, not seeking to return to stable equilibrium under external disturbances and changes due to local dynamics, but adapting and adjusting to changing internal or external processes. Secondly, it has to consider economic, social and ecological heterogeneity by concentrating on the form and the function and process of urban systems (Pickett et al., 2004). Thirdly, resilience planning needs to be based on systems analysis, that will enable authorities to define the points and issues of vulnerability of urban systems and to focus on key issues, being those related to the adaptive and transformative capacities of urban areas in terms of determining strengths and weaknesses in the context of opportunities and threats.

As part of its programme on strengthening urban resilience in African cities, ActionAid commissioned research to better understand the risks faced by urban poor people on the African continent. This exploratory research comprised a desktop review of the literature on urban risk in Africa, and fieldwork in three cities in Senegal, The Gambia and Zimbabwe. It examined hazards, vulnerabilities, local capacities, power imbalances and underlying risk drivers to identify strategies for enhancing resilience to disasters, climate change and conflict in Africa's urban environments. This research shows that disaster risks in towns and cities are strongly linked to underdevelopment. Insecure livelihoods, a lack of basic infrastructure and services, such as water and waste management, poor urban and land planning, inadequate oversight of urban planning, land-use and building standards, and low accountability for the provision of infrastructure and basic services all increase poor people's exposure to hazards, and vulnerability to their effects (Allen et al., 2006; Satterthwaite, 2008; United Nations Human Settlements Programme, 2010). Consequently, reducing risk and building resilience to disasters in urban areas requires tackling the developmental issues that underlie it. This requires improving infrastructure and services, and strengthening livelihoods, all of which are critical in reducing exposure to hazards and enhancing people's ability to cope with and recover from disasters. It is essential to facilitate and support efforts by governments to reduce risk, while at the same time holding them to account through transparent, responsive and proactive governance structures. It is equally important to involve the private sector, as business and industry often contribute to risk on the continent.

Policy frameworks reflect policy objectives and the way in that these will be implemented. Increasing urban climate resilience requires municipal governments to focus on a number of areas, including updating legal and regulatory frameworks to incorporate climate risk considerations; mainstreaming climate change into policy, planning, and budgeting across departments; and increasing capacity to plan, design, and implement climate resilient initiatives. Financing these initiatives is often seen as an additional expense to consider within an already insufficient budget. In some cases, climate resilience initiatives will likely require municipalities to seek funding from national budgets, international donors, or the private sector, particularly when capital expenditure for infrastructure is involved. In other cases, cities may find opportunities to self-finance climate resilience programmes through improved governance of services and budget management, or through innovative public finance options.

Climate-resilience objectives are reflected, to a lesser extent, in legislative and regulatory frameworks. In Kenya, a climate change bill was presented to parliament in 2012 but failed to get presidential endorsement. The government prepared a national climate change response strategy and published a national climate change action plan in March 2013 to implement this strategy (Njoroge et al., 2017). In Gambia, the Programme for Accelerated Growth and Employment (PAGE 2012–2013) identified the need to develop a low carbon development strategy and to mainstream climate (UNFPA, 2017). Tracking Adaptation and Measuring Development (TAMD) was piloting approaches to monitoring and evaluating climate change interventions to support feedback and learning within the planning process (Brooks et al., 2011).

Countries have initiated the process of costing climate resilient activities and integrating these costs into national and sectoral development plans (Ayers & Huq, 2009; Biagini & Miller, 2013). For example, in Bangladesh and Cambodia, a Climate Change Public Expenditure and an Institutional Review (CPEIR) were carried out, respectively, to explore how to integrate climate change issues into expenditure decision-making and responsive fiscal policies. Gambia cost its national climate change action plan and integrated the same into its three-year development plan. Bangladesh also cost its climate change strategy and action plan that guided investments by the climate change trust funds (Am et al., 2013). Rwanda relied on budgetary analysis and financing gaps methodology to determine indicative climate-relevant costs. The T21 model, prepared by Kenya, is likely to support the costing and integration of climate change interventions into national development priorities and budgets. In terms of resource mobilisation, most countries rely on international resources to meet the scale of investment required for climate-resilient development. There are a few exceptions: although the information from these country studies provides some indication of how climate change is being mainstreamed and the utility of the framework, more work still needs to be done on the subject. An evolution toward country-led institutional arrangements would support a more sustainable integration of climate resilience into development planning. Resilience has become an important issue in urban policy (Evans, 2011; Davoudi et al., 2012).

Local land-use planning often is referred to as the "constitution for future development" (Tang & Dessai, 2009: 368) since it encompasses most of the locality's planning area, affects significant development concerns, reflects the community's development goals, and represents the future direction of public policies. Thus, integrating CCA into land-use planning is a key strategy to ensure sustainable development and efficient use of limited resources amid climate change conditions. The approach enhances the capability of communities to address the present and expected climate change risks, and to respond to and recover from climate change impacts (Klein et al., 2005, Klein et al., 2007; IPCC, 2014).

Likewise, mainstreaming CCA into land-use planning is an important task because of the complex relationship between climate change and land-use. Climate change affects both the "demand and supply for space" (Koomen et al., 2008: 262). By altering the present and future use of that space, climate change influences the use, productivity, and access to land. Consequently, land issues and policies are vital concerns in adaptation planning (Quan & Dyer, 2008). Mainstreaming CCA into land-use planning can be through (1) expanding planning horizons to incorporate longer climate predictions; (2) strategising development in flood prone and other high-risk areas; (3) considering the medium- to long-term risks posed by climate change in vulnerable areas; (4) revising land-use regulations and standards that reflect climate variability; and (5) incorporating climate change risk assessments into land information systems, among others (UNDP-UNEP, 2011; IPCC, 2014).

Urban adaptation planning requires planners and policy-makers to co-operate with local governments to be able to undertake necessary actions and strategies (Hamin and Gurran, 2009). This includes revising present planning and building codes and investing in civic education and engagement. A recent survey (ICLEI, 2012) shows that global actions towards urban adaptation planning are being undertaken, ranging from preparatory work to implementation of strategies and plans. However, these actions are mainly dealing with climate-related changes. It is a critical aspect that the contemporary urban adaptation planning is almost solely focused upon environmental and climate change. Climate change is indeed an important and urgent topic that needs to be dealt with but by excluding other vital aspects, such as the societal and urban changes discussed above, the cities that are planned today will not be able to reach a complete state of resilience.

Research Methodology

The study used qualitative and quantitative methods of research. The researcher collected data from both primary and secondary data sources. For primary data, the researcher used focus group discussion, interviews, GIS tools and observation as data collection tools. For secondary data, the researcher used literature review. Purposive sampling was used in selecting key informants of the study who are two planners from Masvingo City Council and from Department of Physical Planning and one Environment Officer from Environmental Management Agency. Observations were made to complement information from interviews. Focus group discussions were employed to compliment information that was given through the interviews. Spatial data was used through importing scanned maps, such as vegetation, geological and weather maps or other maps created in other cartographic tools, such as AutoCAD.

Data Analysis and Interpretation

The city of Masvingo is the oldest colonial urban settlements in Zimbabwe and is the capital city of Masvingo Province. The city attained municipal status in 1953 and city status in 2002. It is located in the south eastern part of Zimbabwe. The administrative district of Masvingo city is strategically positioned at the junction of two national roads (R9 and R4). City of Masvingo is situated in agro ecological region five that receives little rainfall of about 600 mm per annum rendering the area prone to droughts. The town is located at the confluence of two rivers Mucheke and Shagashe River. Small tributaries of the main river intersect the area and form undevelopable wetland areas. These watercourses have an important impact on the development of the city. Masvingo lies over a geological formation of intrusive igneous rocks. Different soil types exist in the area and vary between moderately shallow greyish, brown, coarse grained sands and reddish-brown clay loams. The geology is, generally, suitable for physical development. The natural vegetation of Masvingo is predominantly Savanna, interspersed with grassland and woodland. Lake Mtirikwi and Kyle Recreational Park from the main habitat for flora and fauna species due to their protected nature (Fig. 12.1).

According to World Health Organisation (2017) the population of Masvingo urban is estimated at 161,710. Furthermore, Masvingo City covers an estimated 72.88 square kilometres. The city has since grown from a small town of less than

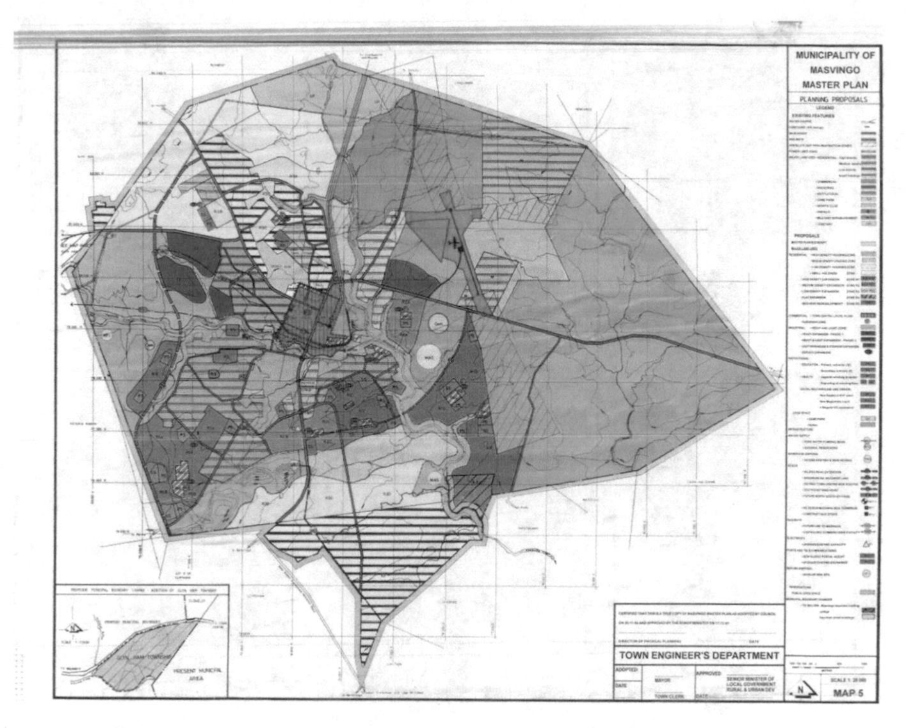

Fig. 12.1 Master plan for Masvingo City. *Source* Department of Physical Planning Masvingo Province (1996)

50,000 inhabitants in the 1990s to a staggering 88,000 urbanites in the year 2012 (ZIMSTATS, 2012). The dominant land-uses that anchor the economy of the city are commerce and the booming residential uses.

The respondents from EMA, MCC and DPP identified land-use functions in Masvingo City namely industry, residential, institutional, recreational, green zones, commercial and supportive infrastructure and agriculture (urban). There has been an increase in built-up area in Masvingo between 1999 and 2019 and corresponding decrease in open spaces.

Figure 12.2 shows that there have been significant changes in land-uses in Masvingo City over time that have a bearing on local climate. The change can be attributed to what Marcuse and Van Kempen (2000) termed as a constant process of internal and external change where cities decline or expand, developing new form and function, dealing with various difficulties, such as segregation, changing demographics and spatial patterns, economic crisis and global competition. In 1999, more area was under vegetation cover in Masvingo City as compared to 2019. Also Fig. 12.2 indicates that housing developments projects have taken place over time. This is seen by the increase in land covered by housing development and built up area. The maps of 1999 up to 2003 show wide green open spaces but as years are passing by the green open spaces are fast disappearing into densely built up areas.

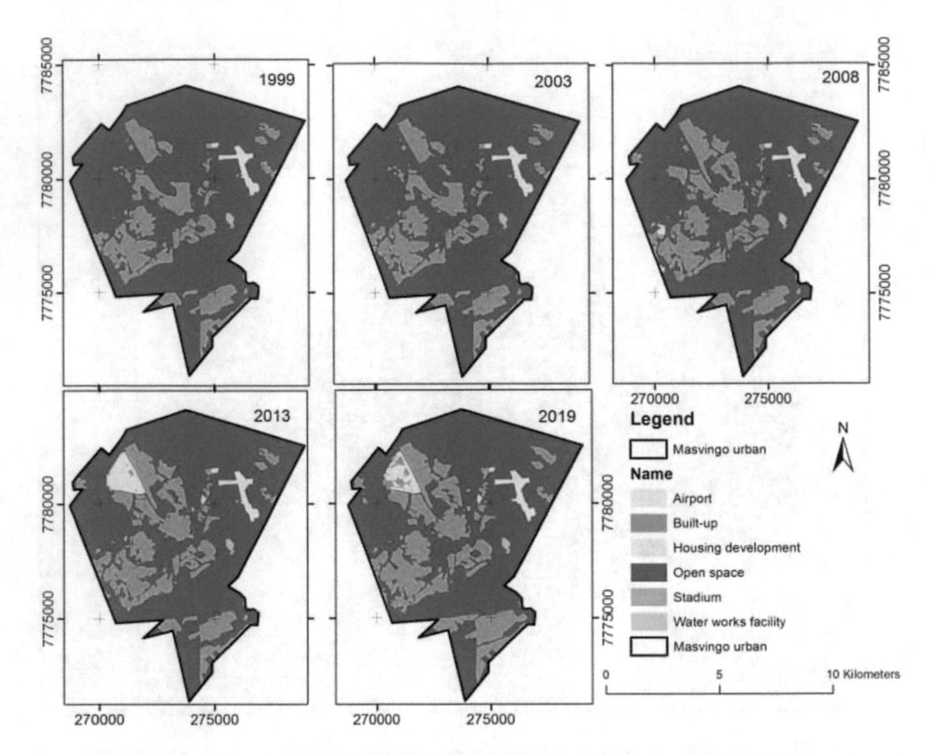

Fig. 12.2 Land-use functions in Masvingo City 1999–2019. *Source* Survey 2019

These areas include Victoria Ranch and Runyararo West that used to be grazing land and paddocks for communal farmers in Masvingo Rural District Council but has fast changed into high-density suburbs less than a decade ago. These developments have come with a reduction in vegetation cover that means more erosion on the bare ground and less trees to act as carbon sinks in carbon sequestration. The changes in area under vegetation cover (open spaces) are also shown in Table 12.1.

Table 12.1 Net change in coverage of land-use functions in Masvingo in square metres

Name	2019	2013	2008	2003	1999
Housing development	1,225,317.341	1,811,512.386	257,665.3	54,470.92	54,470.92
Water works facility	154,299.177	154,367.614	154,367.6	154,367.6	154,367.6
Open space	50,897,294.1	52,916,875.89	55,045,692	56,574,806	56,870,107
Airport	948,945.238	928,347.059	928,347.1	928,347.1	933,018.9
Built-up	18,611,307.88	16,075,667.72	15,500,698	14,176,820	13,876,848
Stadium	39,627.831	24,143.878	24,143.88	22,102.96	22,102.96

Source Survey 2019

The attention for resilience has grown in scientific literature and most literature is about defining the concept and the content of resilience as an approach to understand the outcome of resilient systems and what such systems should look like. Interviewees in Masvingo City were mostly focused on only one principle, 'Encouraging learning' in that the focus was on educating the society to create the need of resilient systems. The respondent from MCC said that;

> Municipalities need a culture of learning, a learning infrastructure and the skills to maximise learning opportunities. Planners are constrained by the bureaucratic mind set that rewards stability and conformity more than learning and innovation. Though they have heard of resilience, they cannot apply it for fear of mistakes leading to disregard by the system as petty planners. The Masvingo landscape does not suit the principles as directives come from the government and bosses from the city council.

In the same line of thinking, according to Reed et al. (2013), learning processes are important for gaining new knowledge of its functions and vulnerabilities and initiate collaborative action that may lead to change. The capacity to learn is cited as a key characteristic of resilience and forms a central aspect of research and practice on resilience in socio-ecological systems (Reed et al., 2013). This is also necessary to exchange knowledge for the development of policies within cities.

The respondent from EMA noted that urban agriculture that is being practised in Masvingo does not meet the requirements of the principles of resilience to climate change as most of it is stream bank cultivation along Mucheke and Mushagashe rivers and there are no conservation works practised. He also went on to note that most industries pollute the environment. This is evidenced by the water hyacinth in Mucheke and Shagashe River.

Housing developments do not meet the principles of climate change as they worsen the situation by clearing vegetation or trees that usually act as climate regulators, for example, in Victoria Range that was once a densely vegetated area but is now bare and dusty because of the ongoing deforestation. The deforestation is as a result of people clearing land for residential developments and fetching firewood since the area is not yet connected to any electricity grid. The high dependence of these cities on the surrounding natural environment amplifies the risks of environmental degradation (White & Engelen, 1997). Robichaud and Wade (2011) concurs with this as he noted that urban expansion alters land surfaces and disrupts natural processes and systems, aggravating flooding, for example, by covering ground with hard surfaces, such as roofs, roads and pavements that do not absorb rainwater as vegetation does, increasing run-off and pooling. Construction has also destroyed wetlands that aid climate change resilience. As had been earlier noted by Cross (2001), the location, geography, and size of settlements, the built environment and infrastructure, the natural resources at risk, and sea level rise are important factors contributing to biophysical vulnerability. Hence, the developments happening in Masvingo City contribute to higher risks in terms of disasters.

The respondent from MCC noted the reasons contributing to this lack of resilience are caused by the lack of the government to plan with a green face. She said,

> For many decision-makers the environment remains a luxury good, deserving of attention and budget resources only once more pressing needs with regards to housing and basic

> services have been satisfied. It is a view that fails to acknowledge the important linkages between human well-being, development and the health of the natural environment and this has hindered many local authorities to focus on resilience but on providing shelter for all as mandated by the government.

The respondent from DPP was of the view that these land-uses meet the principles of resilience to climate change to a limited extent because the city lacks a positive robust environment in its design hence its limited capacity to anticipate and respond to potential issues. Civil engineering designs for city infrastructure lack modernity and capacity to absorb shocks of climate change. Infrastructure is not adaptive and may encourage negative response to the effects of climate change, for instance, the houses in the *kuma* '*R*' area that have long suffered urban decay. Such areas are prone to flooding. This redundancy of infrastructure militates quick recovery in situations requiring immediate response to climate change.

Moreover, it is observed that green spaces are fast disappearing as more open spaces are being converted to other uses, such as residential. This therefore, means that there are fewer urban ecosystems and green infrastructure. This has been attributed to lack of resourcefulness or limited citizen empowerment as citizens and institutions have limited knowledge of climate risks, and are viable to adapt to shocks and stresses in the changing environment. From the above findings, it can be noted that land-uses in Masvingo City have changed over time with more land being cleared for new housing developments that then translated to less vegetation cover meaning that is vital for carbon sequestration. The responded from the DPP said,

> The Department of Physical Planning needs to have framework and approaches that can be used by residents to cope with housing change and these are elements of adaptability, collective resilience in overall form and structure, examining vertical and horizontal housing adaptation and change and reviews adaptation in alleyways and public space.

Table 12.1 indicates that the built-up land cover in Masvingo City has been increasing since 1999. Open spaces have dwindled in coverage due to housing developments that have increased. The responded from MCC noted that this has led to deteriorating environmental trends in the form of high population density initiating unsustainable land-use practises. The respondent from EMA also said that this rapid increase in clearing land for housing development has a negative effect on the environmental processes and quality. The data also revealed an increase in the built-up area of Masvingo City and consequently, a decrease in the area covered by vegetation. In the year 1999, built up area covered 13,876,848 m^2 while, by the year 2019 it had almost doubled.

The temperature trends for Masvingo City from 1952 to 2016 indicate an increase in temperature over the years. The respondents' feedbacks were analysed based what they observed over the time they lived in their communities. Members of the MRA confirmed that winter temperatures are apparently becoming warmer and summer temperatures are characterised by dry spells and heat waves. The general responses indicate that the temperature patterns are now erratic observing increasing ambient temperatures, especially in summer, that are responsible for high evapotranspiration rates.

Figure 12.3 shows rainfall deviation for Masvingo City from 1998 to 2018. The graph shows that rainfall in Masvingo since 1998 has been deviating more from mean in a negative way with an increased magnitude. All respondents perceived climate change to be taking place. The respondent from MCC said that the onset of the rainfall season is now coming late and cessation time coming earlier than before with some mid-season droughts in between. The cessation of rainfall is also unpredictable with rainfall periods extending into June and July and sometimes ending much earlier in March. The frequency and magnitude of flooding has also increased in the Masvingo CBD. This shows the need for landscape resilience.

The respondents from the Masvingo Residents Association were in agreement that they are noticing changing urban temperatures or rising temperatures and increasing threats of water shortages due to changes in rainfall patterns and quantities. This has distorted urban landscapes in the sense that people are now digging wells on their homes or in their communities and when one well ceases to have water, they dig another one elsewhere leaving holes all over. These holes end up being a danger to the people when they walk at night and the holes collect water in the rainy season creating stagnant water bodies. The respondent form EMA also noted water shortages in Masvingo as effects of climate change and he said,

> Rapid urbanisation is, generally, putting high pressures on limited urban resources, like fresh water, while at the same time producing large amounts of waste water and wastes. The urban demand of fresh water is rising rapidly while the availability of fresh water is becoming a serious problem. The MCC lacks resilient water infrastructure that can sustain the city through the droughts that are happening each year as there are more water leaks.

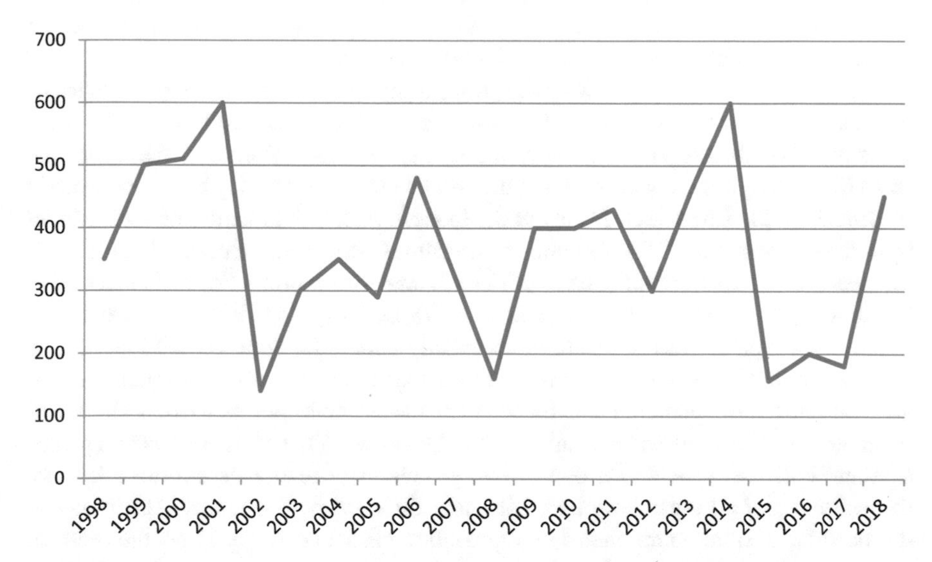

Fig. 12.3 Annual rainfall deviations for Masvingo city from 1998–2018. (Adapted from Chikodzi 2018)

Koomen et al. (2008) noted that climate change is aggravating existing risks and creating new ones. The effects of climate change in Africa are as varied as its diverse environments and climates, but include changing temperatures, more intense storms, variable and unpredictable weather and increased occurrence of drought are the predominant ones. Also, any factor that makes an impression on the urban landscape, such as climate change, will affect people's lives directly or indirectly. The impacts of climate change are likely to undermine planned development outcomes posing significant challenges to the resilience of livelihoods and ecosystems, for instance, the already noticeable indirect effects on urban habitants and diversity through changes in rainfall, rising temperatures and extreme events.

According to the informant from Department of Physical Planning, there are no policies earmarked for resilience to climate change, what exists is the consciousness of the planning discipline to the process. He further argued and said,

> Resilience is encompassed in strategic plans (master, local and layout plans) but it is not mandated by law but merely a case of planners being conscious of the need to incorporate climate issues in their medium to long term plans.

However, the respondent from EMA was of the view that urban developments are guided by statutes, such as the Environmental Management Act (20: 27), Forestry Act and the Water Act. This shows that there seems to be lack of coordination between the two bodies as to how they can work together to ensure a robust and resilient modern Masvingo City. As noted by Jabareen (2012), urban resilience refers to learning, planning, forecasting, resisting, absorbing, accommodating to and recovering from unforeseen changes within cities. Since city resilience is a complex and multidisciplinary phenomenon, it requires cooperation between various actors working within governance, spatial, economic and social urban dimensions.

Davoudi (2012) concurred with the above as he argued that the characteristics of agents, institutions, and systems that contribute to resilience are likely to be most effective when they work in synergy, rather than isolation. Moreover, the informant from Masvingo City Council said that they conduct their day to day business strongly informed by the EMA Act, Mines and Minerals Act, Urban Councils Act, Model building by-laws, the RTCP Act and other statutes. However, he noted that there are no clear provisions for resilience to climate change or adaption. For instance, model building by-laws provide that there is need for building lines when constructing a house. However, it does not state what exactly should be done on the land that is left. As a result, people end up paving those spaces yet if it was stipulated that the area be must have plants and gardens, then these would act as carbon sinks and promote percolation of water to underground sources. What exist are bits and pieces of statements that provide for one or two principles of resilience but not all. Also, the existence of all these laws that fall under different bodies is clear testimony of the need for a more comprehensive approach to climate change. Thus, the relevant sections of the RTCP Act should make sufficient reference to the subsisting law dealing with environmental matters, that is integrate the EMA Act into the RTCP Act.

The respondent from EMA noted that Section 59-100 of the Environmental Management Act provides for urban landscape resilience. It states that it is the duty of EMA to regulate and monitor the management and utilisation of ecologically fragile ecosystem to make model by-laws and to establish measures for the management of the environment within the jurisdiction of the local authorities. It also provides for the development and implementation of incentives for the protection of the environment and regulation and monitoring of access by any person to the biological and genetic resources of Zimbabwe. This means that they provided for sustainable development of cities in such a way that they uphold urban resilience principles. However, although such a statute exists, not much enforcement can be noted since there is evidence of deforestation as people cut down trees for firewood in existing green spaces in Masvingo. There is also sand extraction occurring in areas, such as Victoria Ranch and Morningside and along Mucheke River.

Furthermore, the respondent from EMA said that Masvingo City stands guided by the Water Act (1998) that was signed into law after considerable consultation with stakeholders. The act is founded on economic efficiency, environmental sustainability and equity of use. Water can no longer be privately owned. Water is to be viewed from the complete hydrological perspective that is groundwater and surface water are treated as part of one hydrological system.

There is greater consideration of the environment, with environmental water use now recognised as a legitimate user. There is more control over pollution, with the "polluter pays" principle being introduced. However, on the ground, it can be noted that not much enforcement of the law is being done as the water bodies in Masvingo, such as Mucheke River and Mushagashe are severely polluted by industries and sewage. Insecure livelihoods; a lack of basic infrastructure and services, such as water and waste management; poor urban and land planning; inadequate oversight of urban planning, land-use and building standards; and low accountability for the provision of infrastructure and basic services all increase poor people's exposure to hazards, and vulnerability to their effects.

A lot of suggestions were made by representatives from Masvingo Residents Association focus group discussion, the officials from DPP, EMA and Masvingo City Council. Many researchers have concluded that cities also hold the possibilities and opportunities in becoming hubs for positive change and transformation, that includes finding new solutions and innovations in striving to become increasingly resilient (Olazabal et al., 2012). The key recommendations from theses stakeholders were that there is need for proactive, rather than reactive approaches to manage climate change and to stiffen penalties and continuous review of the available acts on climate issues. It was also suggested that there was need for expanding or embracing environmental aspects of planning to specifically require consideration of the need to adapt to climate change in city planning and budgetary processes.

Respondents from EMA noted that there was need for Masvingo City Council to promote higher resilience of the natural systems and take advantage of the lower costs associated with having these. They also noted that the Council should be reinventing the energy system for powering the city. It was noted by the official from DPP that it was high time for Masvingo City Council to have robust policies and infrastructure provisions that enable the city to be resilient to shocks and disturbances. Masvingo Residents Association also observed that there is need to reduce the use of concrete surfaces in building cities, such as paved streets and ad parking lots that do not allow water to percolate thereby increasing chances of flooding since there will be more runoff. Of which Müller (2011) argued that to be resilient to disasters, therefore, means to be flexible in coping with such events and be able to make the best of the opportunities that might arise. Similarly, climate act planning for Chicago" from the Chicago Ministry of Environment (2008) proposed adaptive strategies, including mitigating urban heat islands, implementing city-based innovations, improving air pollution management, managing rainstorms, establishing green design, protecting trees, and providing an incentive to the public and enterprises to actively respond to climate change.

Conclusion and Recommendations

The research examined the landscape resilience to climate change in Masvingo City and it found out that the existing urban landscape in Masvingo City does not meet the principles of resilience. Resilience is still evolving as a concept for sustainable development. The study made some recommendations for policy, research and practice. Evidence from the research indicates that the existing urban landscape in Masvingo City does not meet the principles of resilience. Resilience is still evolving as a concept for sustainable development. For most interviewees, the definition of resilience was vague and also the approach for using the concept as a tool was unclear. The study revealed that the impacts of climate change are likely to undermine planned development outcomes posing significant challenges to the resilience of livelihoods and ecosystems. For instance, the already noticeable indirect effects on urban habitants and diversity through changes in rainfall, rising temperatures and extreme events.

There is lack of policies governing the provision for urban landscape resilience in face of climate change in Masvingo City, lack of integration between different resilient and adaptation policies at various scales, and the limited incorporation of resilience and adaptation as planning concepts. There is also lack of policy coordination in Masvingo City regarding climate change resilience. Governing institutions do not have policies that directly govern issues to do with climate change but are only guided by planning tools that are related to climate change resilience that are zoning regulations, subdivision and setbacks, building standards, for example, minimum floor level, avoiding human occupancy of hazard zone and hazard mapping. This clearly shows that coverage of climate change resilience and adaptation is limited in Masvingo City and a few policies have strategic plans to enhance resilience to climate change. There are no resources allocated towards the enhancement of resilience to climate change. This is a widespread pattern in many developed and developing countries as reviewed.

The study found out that EMA; that is meant to coordinate the web of cross cutting initiatives, lacks political power, budget and institutional capacity. Climate change adaptation policies managed by EMA tend to have more of a rural and less of an urban focus. Even when EMA comes up with urban policies, it focuses primarily on the prevention and mitigation of natural disasters provoked by extreme weather events and not the resilience part to climate change.

It can be concluded that urban landscape resilience efforts to climate change in Masvingo City are lagging behind in contributing to long term sustainability of the city. The study revealed that respondents are acknowledging problems they are observing in climatic conditions and in changes in their surrounding environment, but the resilience part is still a new concept that needs to be mainstreamed in their daily activities. It was revealed that the concept of resilience is quite plastic and is sometimes loosely equated with reducing vulnerability or enhancing adaptive capacity. As resilience becomes mainstreamed into efforts to acclimatise development, there is a need for continued questioning of how the concept is used and applied to urban areas.

The study recommends for:

- An urban policy that guides government and local governments on climate change and resilience to climate change. It is essential that climate change be fully integrated into development planning.
- Incorporating climate adaptation and resilience into professional education that is crucial for building resilience. More effective engagement of planners in adaptation can influence attitudes and behaviour of other agents, especially the community, local government, and development industry, and contribute to resilience.
- Resource mobilisation towards funding climate change resilience and adaptation by the government and local authorities. High priority should be towards channelling resources to local community level institutions to develop resilience. Elements of this can support the strategies people in Masvingo are already employing at the household level to strengthen the robustness of key assets (e.g. housing) in the face of worsening patterns of severe weather. Similar actions can be identified for small businesses and at the community level to protect urban assets, such as drainage systems, recreation spaces, schools or sanitation facilities.
- Participatory planning of all stakeholders from the grassroots level up. Processes of participatory planning and deliberation addressing climate change trends and impacts can help to strengthen and focus collective action to ensure that resilience is built through better local level land-use.
- The MCC, DPP, EMA and other related sectors to carry out coordinated land-use planning. To support landscape restoration, cross-sectoral coordination is essential. Agencies often work in relative isolation, and even at cross-purposes. This is at least partially due to the institutional structure and the lack of capacity of these institutions to cooperate closely in land-use planning and management. There is a need for institutions dealing with ecosystem and land-use issues to integrate the management of natural resources (in particular, forests, trees, soil and water) through improved, multi-sectoral land-use. National institutions, policies and laws need to support actions to build resilience in every sector at local level. Once an

understanding is reached of the needs related to building resilience, the institutional framework for urban landscapes and related sectors should be reviewed to see where adjustments are needed for the support of efforts to build resilience.

- The implementation of new urbanism by all stakeholders directly involved in urban planning, like urban councils and DPPs that advocates for smart resilient cities. Integrated Development Planning (IDP) and the new urban planning system have been conceived under the premise that traditional hierarchical governments with agencies separated into silos might not be the most accurate form of governance when sector policy and decisions have all an impact on the space. Hence, new urbanism offers the realm for the coordination and integration of different government sectors.

References

Ahern, J. (2011). Green infrastructure for cities: The spatial dimension. In V. Novotny, & P. Brown, (Ed.), *cities of the future: Towards integrated sustainable water and landscape management*. IWA Publishing.

Alberti, M. (2003). *Advances in Urban ecology: Integrating humans and ecological processes in Urban ecosystems*. Springer.

Allen, A., Dávila, J. D., & Hofmann, P. (2006). The peri-urban water poor: Citizens or consumers? *Environment and Urbanization, 18*(2), 333–351.

Am, P., Cuccillato, E., Nkem, J., & Chevillard, J. (2013). Mainstreaming climate change resilience into development planning in Cambodia. Available online: https://www.gcca.eu/sites/default/files/soraya.khosravi/a6_-_10047iied.pdf

Arefi, M. (2011). Order in informal settlements: A case study of Pinar, Istanbul. *Built Environment, 37*(1), 42–56.

Ayers, J. M., & Huq, S. (2009). Supporting adaptation to climate change: What role for official development assistance? *Development Policy Review, 27*(6), 675–692.

Batty, M. (2013). Polynucleated Urban landscapes. *Urban Studies, 38*(4), 635–655.

Bathelt, H., Malmberg, A., & Maskell, P. (2004). Clusters and knowledge: Local buzz, global pipelines and the process of knowledge creation. *Progress in Human Geography, 28*(1), 31–56.

Berkes, F. (2007). Understanding uncertainty and reducing vulnerability: Lessons from resilience thinking. *Natural Hazards, 41*, 283–295. https://doi.org/10.1007/s11069-006-9036-7

Biagini, B., & Miller, A. (2013). Engaging the private sector in adaptation to climate change in developing countries: Importance, status, and challenges. *Climate and Development, 5*(3), 242–252.

Brand, F. S., & Jax, K. (2007). Focusing the meaning(s) of resilience: Resilience as a descriptive concept and a boundary object. *Ecology and Society, 12*(1), 23–39.

Brooks, D. M., Pando-V. L., Ocmin-P. A., & Tejada-R. J. (2011). Resource separation in a napo-amazonian gamebird community. In: D. M. Brooks, & F. Gonzalez-F. (Eds.), *Biology and conservation of cracids in the New Millenium*, (pp. 213–225, No. 2) Miscellaneous Publications of the Houston Museum of Natural Science.

Bruneau, M. (2003). A framework to quantitatively assess and enhance the seismic resilience of communities. *EERI Spectra Journal, 19*(4), 733–752.

Burby, R. J. (2009). *Cooperating with nature: Confronting natural hazards with land use planning for sustainable communities*. Joseph Henry Press.

Chang, S. (2009). *Conceptual framework of resilience for physical, financial, human, and natural capital*. School of Community and Regional Planning, University of British Columbia.

Chelleri, L., & Olazabal, M. (2015). Findings and final remarks. In L. Chelleri, & M. Olazabal, (Ed.), *Multidisciplinary perspectives on urban resilience: A workshop report*. Basque Centre for Climate Change.

Chisi, et al. (2000). *Research project guide*. The Zimbabwe Open University.

Coaffee, J. (2008). Risk, resilience, and environmentally sustainable cities. *Energy Policy, Elsevier, 36*(12), 4633–4638.

Cross, N. (2001). Designerly ways of knowing: Design discipline versus design science. *Design Issues, 17*(3), 49-55.

Cutter, S. L., Barnes, L., Berry, M., Burton, C., Evans, E., Tate, E., & Webb, J. (2008). A place-based model for understanding community resilience to natural disasters. *Global Environmental Change, 18*(4), 598–606.

Davidson, J. L., Jacobson, C., Lyth, A., Dedekorkut-Howes, A., Baldwin, C. L., Ellison, J. C., Holbrook, N. J., Howes, M. J., Serrao-Neumann, S., Singh-Peterson, L., & Smith, T. F. (2016). Interrogating resilience: Toward a typology to improve its operationalization. *Ecology and Society, 21*(2), 27–41.

Davoudi, S. (2012). Resilience: A bridging concept or a dead end? *Planning Theory and Practice, 13*, 299–307.

DCCEE. (2011). The critical decade: Climate science, risks and responses. Canberra: Commonwealth of Australia. Department of Health. (2006). A Green Hospital. https://www.newrch.vic.gov.au/Agreenhospital.

Desouza, K. C., & Flanery, T. H. (2013). Designing, planning, and managing resilient cities: A conceptual framework. *Cities, 35*:89–99. https://doi.org/10.1016/J.Cities.2013.06.003

Elmqvist, T., Barnett, G., & Wilkinson, C. (2014). Exploring Urban sustainability and resilience. In L. J. Pearson, P. W. Newton, & P. Roberts (Eds.), *Resilient sustainable cities: A future* (pp. 19–28). Routledge.

Ernstson, H., Van Der Leeuw, S. E., Redman, C. L., Meffert, D. J., Davis, G., Alfsen, C., & Elmqvist, T. (2010). Urban transitions: On Urban resilience and human-dominated ecosystems. *Ambio, 39*(8), 531–545.

Evans-Cowley, J. S., & Gough, M. Z. (2011). Evaluating environmental protection in Posthurricane Katrina plans in Mississippi. *Journal of Environmental Planning and Management, 51*(3), 399–419.

Folke, C., Carpenter, S., Elmqvist, T., Gunderson, L., Holling, C. S., & Walker, B. (2002). Resilience and sustainable development: building adaptive capacity in a world of transformations. *AMBIO: A Journal of The Human Environment*, *31*(5), 437–440.

Folke, C. (2006). Resilience: The emergence of a perspective for social–ecological systems analyses. *Global Environmental Change, 16*(3), 253–267.

Gallopin, G. C. (2006). Linkages between vulnerability, resilience, and adaptive capacity. *Global Environmental Change, 16*, 293–303.

Godschalk, D. R., Brody, S., & Burby, R. (2003). Public Participation in natural hazard mitigation policy formation: Challenges for comprehensive planning. *Journal of Environmental Planning and Management, 46*(5), 733–754.

Hamin, E. M., & Gurran, N. (2009). Urban form and climate change: Balancing adaptation and mitigation in the US and Australia. *Habitat International, 33(3)*, 238–245.

Hsieh, C. H., & Feng, C. M. (2014). Road network vulnerability assessment based on fragile factor interdependencies in spatial-functional perspectives. *Environment and Planning A, 46*(3), 700–714.

ICLEI. (2012). Local governments for sustainability. Available online: https://www.cbd.int/doc/nbsap/nbsaprw-clamed-01/bsap-montpellier-iclei-connect-europe.pdf

IPCC. (2012). Managing the risks of extreme events and disasters. Available online: https://www.ipcc.ch/report/managing-the-risks-of-extreme-events-and-disasters-to-advance-climate-change-adaptation/.

IPCC. (2014). Climate change 2014. Synthesis report. Available online: https://archive.ipcc.ch/pdf/assessment-report/ar5/syr/AR5_SYR_FINAL_Front_matters.pdf.

IPCC. (2007). Climate Change 2007: Appendix to synthesis report. In A. P. M. Baede, P. Van Der Linden, & A. Verbruggen (Eds.), *Climate Change 2007: Synthesis Report. Contribution of Working Groups I, II and III to the Fourth Assessment Report of the Intergovernmental Panel on Climate Change* (pp. 76–89). Geneva: Author.

Jabreen, Y. R. (2012). Sustainable Urban forms: Their typologies, models, and concepts. *Journal of Planning Education and Researc, 26*, 38–52.

Keiner, M., & Kim, A. (2007). Transnational city networks for sustainability. *European Planning Studies, 15*(10), 1369–1395.

Klein, R. J. T., Eriksen, S. E. H., Naess, L. O., Hammill, A., Tanner, T. M., Robledo, C., & O'Brien, K. L. (2007). Portfolio screening to support the mainstreaming of adaptation to climate change into development assistance. *Climatic Change, 84*, 23–44.

Klein, R. J. T., Nicholls, R. J., & Thomalia, F. (2003). Resilience to natural hazards: How useful is this concept? *Global Environmental Change Part b: Environmental Hazards, 5*(1–2), 35–45. https://doi.org/10.1016/J.Hazards.2004.02.001

Klein, R. J., Schipper, E. L. F., & Dessai, S. (2005). Integrating mitigation and adaptation into climate and development policy: Three research questions. *Environmental Science & Policy, 8*(6), 579–588.

Koomen, E., Loonen, W., & Hilferink, M. (2008). *Climate-change adaptations in land-use planning; a scenario-based approach.* In L. Bernard, A. Friis-Christensen, & H. Pundt (Eds.), *The European information society taking geoinformation science one step further: Lecture notes in geoinformation and cartography series.* Springer-Verlag Berlin Heidelberg.

Lankao, P. R. (2010). Water in Mexico city: What will climate change bring to its history of water-related hazards and vulnerabilities? *Environment and Urbanization, 22*(1), 157–178.

Lankao, P. R., & Qin, H. (2011). Conceptualizing urban vulnerability to global climate and environmental change. *Current Opinion in Environmental Sustainability, 3(3)*, 142–149.

Leichenko, R. (2011). Climate change and Urban resilience. *Current Opinion in Environmental Sustainability, 3*(3), 164–168.

Linkov, I., Bridges, T., Creutzig, F., Decker, J., Fox-Lent, C., Kröger, W., & Nyer, R. (2014). Changing the resilience paradigm. *Nature Climate Change, 4*(6), 407–409.

Lu, P., & Stead, D. (2013). Understanding the notion of resilience in spatial planning: A case study of rotterdam, the Netherlands. *Cities, 35*, 200–212.

Ludwig, D., Walker, B., & Holling, C. S. (1997). Sustainability, stability, and resilience. *Conservation Ecology*. Available From: http://www.consecol. org/Vol1/Iss1/Art7/

Marcuse, P., & Kempen, R. V. (2000). Conclusion: A changed spatial order. In P. Marcuse & R. Van Kempen (Eds.)., *Globalizing cities: A new spatial order?*London: Wiley.

Magis, K. (2010). Community resilience: An indicator of social sustainability. *Society and Natural Resources, 23*(5), 401–416.

Manyena, S. B., O'Brien, G., O'Keefe, P., & Rose, J. (2011). Disaster resilience: A bounce back or bounce forward ability. *Local Environment, 16*(5), 417–424.

Markard, J., Raven, R., & Truffer, B. (2012). Sustainability transitions: An emerging field of research and its prospects. *Research Policy, 41*(6), 955–967.

Meerow, S., Newell, J. P., & Stults, M. (2016). Defining urban resilience: A review. *Landscape and Urban Planning, 147*, 38–49.

Mileti, D. (1999). *Disasters by design: A reassessment of natural hazards in the United States*. New Jersey: Joseph Henry Press.

Muller, M. (2011). Adapting to climate change. *Environment and Urbanization, 19*, 99.

Njoroge, J. M., Ratter, B. M., & Atieno, L. (2017). Climate change policy-making process in Kenya: Deliberative inclusionary processes in play. *International Journal of Climate Change Strategies and Management, 9*(4), 535–554.

Olazabal, M., Chelleri, L., Waters, J. J., & Kunath, A. (2012). Urban resilience: *Towards an integrated approach.* Paper presented at 1st International Conference on Urban Sustainability & Resilience, London, UK. Available online: https://www.preventionweb.net/events/view/28308?id=28308.

Pickett, S. T. A., Cadenasso, M. L., & Grove, J. M. (2004). Resilient cities: meaning, models, and metaphor for integrating the ecological, socio-economic and planning realms. *Landscape and Urban Planning, 69*(1), 369–384.

Preston, B. L., & Stafford-Smith, M. (2009). *Framing vulnerability and adaptive capacity assessment*: Discussion paper. Aspendale: CSIRO Climate Adaptation National Research Flagship.

Preston, B. L. (2013). The climate adaptation frontier. *Sustainability, 5*(2), 1011–1035.

Quan, J., & Dyer, N. (2008). Climate change and land tenure: The implications for land tenure and land policy land tenure. *Working Paper 2. Rome: Food and Agriculture Organisation.*

Raco, M., & Street, E. (2012). Development in London and Hong Kong: Resilience planning, economic change and the politics of post-recession. *Urban Studies, 49*(5), 1065–1087.

Reed, M. S., Podesta, G., Fazey, I., Geeson, N., Hessel, R., Hubacek, K., Letson, D., Nainggolan, D., Prell, C., Rickenbach, M. G., & Ritsema, C. (2013). Combining analytical frameworks to assess livelihood vulnerability to climate change and analyse adaptation options. *Ecological Economics, 94,* 66–77.

Robichaud, I., & Wade, T. (2011). *Planning and policy in Atlantic Canada.* Atlantic Climate Solutions Association.

Satterthwaite, D. (2008). Climate change and urbanization: Effects and implications for Urban governance. In *United Nations Expert Group Meeting on Population Distribution, Urbanization, Internal Migration and Development* (pp. 21–23). DESA.

Simonsen, S. H., Biggs, R., Schluter, M., Schoon, M., Bohensky, E., Cundill, G., & Quinlan, A. (2014). *Applying resilience thinking: Seven principles for building resilience in social-ecological systems.* Stockholm University.

Tang, S., & Dessai, S. (2012). Usable science? The UK climate projections 2009 and decision support for adaptation planning. *Weather, Climate, and Society, 4*(4), 300–313.

Tierney, K. (2007). Disaster governance: social, political, and economic dimensions. *Annual Review of Environment and Resources, 37*, 341–363.

Timmermans, W.; Ónega López, F., Roggema, R. (1981). Complexity theory, spatial planning and adaptation to climate change. In R. Roggema, (Ed.). *Swarming landscapes: The art of designing for climate adaptation* (pp. 43–65) Springer.

Turner, B. L. (1990). Vulnerability and resilience: Coalescing or paralleling approaches for sustainability science? *Global Environmental Change, 20*, 570–576.

Tyler, S., & Reed, S. O. (2011). Results of resilience planning. In: M. Moench, S. Tyler, & J. Lage (Eds.), *Catalyzing Urban climate resilience: Applying resilience concepts to planning practice in the ACCCRN program 2009–2011* (pp. 239–270). ISET.

UNDP. (2011). *Human development report 2010—The real wealth of nations: Pathways to human development*. Palgrave, Macmillan.

Un-Habitat. (2010). Solid waste management in the world's cities. Nairobi: UN-HABITAT.

UNFCC. (2013) National adaptation programmes of action. http://unfccc.int/national_reports/napa/Items/2719.php. Accessed 3 Dec 2013.

UNFCCC. (2013). Report of the Conference of the Parties on its Eighteenth Session, Held in Doha from 26 November to 8 December 2012 FCCC/CP/2012/8/.

UNFPA. (2017). Programme for accelerated growth and employment 2012–15. Available online: https://gambia.unfpa.org/en/publications/programme-accelerated-growth-and-employment-2012-15

UNISDR. (2012). *How to make cities more resilient: A handbook for local government leaders.* Author.

United nations human settlements programme. (2010). *The State of African cities 2010: Governance, inequality and urban Land markets.* UN-HABITAT.

United Nations Human Settlements Programme. (2010). *Cities and climate change: Global report on human settlements* 2011. UN-Habitat.

Vale, L. J., & Campanella, T. J. (2005). Axioms of Resilience. In L. J. Vale & T. J. Campanella (Eds.), *The Resilient city: How modern cities recover from disaster: How modern cities recover from disaster* (pp. 335–355). Oxford University Press.

Walker, B., Gunderson, L., Kinzig, A., Folke, C., Carpenter, S., & Schultz, L. (2006). A handful of heuristics and some propositions for understanding resilience in social-ecological systems. *Ecology and Society, 11*(1), 1–12.

Walker, B., Holling, C. S., Carpenter, S. R., & Kinzig, A. (2004). Resilience, adaptability and transformability in social-ecological systems. *Ecology and Society, 9*(2), 5–8.

Walmsler, A. (2014). Greenways and the making of Urban form. *Landscape and Urban Planning, 33*, 81–127.

Weingart, P., Engels, A., & Pansegrau, P. (2000). Risks of communication: Discourses on climate change in science, politics, and the mass media. *Public Understanding of Science, 9*(3), 261–283.

White, R., & Engelen, G. (1997). Cellular automata as the basis of integrated dynamic regional modelling. *Environment and Planning B: Planning and Design, 24*(2), 235–246.

World Bank. (2011). *Climate change, disaster risks and the urban poor: Cities building resilience for a changing world*. International Bank for Reconstruction and Development.

ZIMSTATS. (2012). National Census Report: Harare: ZIMSTATS.

Chapter 13
Towards a Resilience Framework for Urban Zimbabwe

Patience Mukuzunga, Christine R. Chivandire, and Innocent Chirisa

Abstract The study seeks to develop a resilience framework that is for sustainable urban cities of Zimbabwe. The urban areas of Zimbabwe are riddled with many problems that include urban sprawl, shortage of water, poor service delivery and urban poverty, among others. These problems have degraded the expected standards of the urban areas and hence steps need to be taken to eliminate them and transform urban areas into world first class cities. The concept of resilience framework is a key phenomenon in the 21th century and has been implemented in most developed countries, such as the United States of America and many European countries. Developing countries have also started to adopt this idea in an attempt to develop their countries and cities. The study was done by reviewing the already existing secondary records and a number of case studies were analysed and debated in this paper. The framework will design solutions that will lead to improvement of the quality of living standards, services delivery and a high GDP per capital and clean urban areas. It can be argued that the resilience framework is capable of handling the situation that is being faced by urban areas of Zimbabwe and may give solutions that are sustainable. Its formulation with regard to the situation of Zimbabwe's urban areas is very crucial. Therefore, the formulation of the resilience framework for urban Zimbabwe will lead to the creation of the expected qualities of urban cities. It is recommended that major steps should be taken to bring successful implementation of this framework in shaping the urban areas of Zimbabwe.

P. Mukuzunga · C. R. Chivandire · I. Chirisa (✉)
Department of Demography Settlement and Development, University of Zimbabwe, PO Box MP167, Mt Pleasant, Harare, Zimbabwe

I. Chirisa
Natural and Agricultural Sciences, PO Box/Posbus 339, Bloemfontein 9300, Republic of South Africa

I. Chirisa and A. Chigudu (eds.), *Resilience and Sustainability in Urban Africa*, Advances in 21st Century Human Settlements,
https://doi.org/10.1007/978-981-16-3288-4_13

Introduction

Urban areas of Zimbabwe are in a deteriorating state as they are suffering from high rate of unemployment, pollution, overcrowding, urban sprawl and poverty. Hence there is need for a quick action to resolve this current situation. Many problems identified in these areas have degraded their value and eminence. Zimbabwe's urban areas are one of the worst places in the world and their status has to be reformed. This study ought to identify the problems that are affecting the urban areas of Zimbabwe and to explain how they have led to the formulation of this resilience framework. This paper also explains how best this framework can mitigate and turn the urban areas into the desired cities of the world. Different sources were used to acquire information about the situation in these areas. Most of the data used was gathered from studies that were already done in the past pertaining urban environments of Zimbabwe. The results show the problems that are currently being faced in the urban areas. The study hopes to provide a resilience framework that will tackle the development of these areas. It can be argued that to reach sustainability, resilience should call for a total change in the operating system that is currently occurring among the urban cities of Zimbabwe at the present moment.

This study focuses on the urban areas of Zimbabwe that includes Harare, Bulawayo, Masvingo, Mutare, Gweru and Kwekwe, among others. These are the cities with core problems that are in need of action hence the call for resilience to change their system of operation and to transform them into the world's preferable smart cities with the correct values for a prime city. The word "resilience" simply means to adapt to the events of life. It is an ability to bounce back after a shock or stress. This means that resilience is done as a remedy for a disaster or something that has disturbed the way of doing things that have been prevailing in a certain structure. It is argued that urban resilience is the ability of an urban-system and all its constituent, socio-ecological and socio-technical networks across temporal and spatial scales to maintain and rapidly return to its desired functions in the face of disturbance. It is the ability to adapt to change and quickly transform systems that limit current or future adaptive capacity (Meerow et al., 2015). This means that transformation is the basic element of urban resilience. Hamilton (2009) defines resilience in urban areas as the ability to recover and continue to provide main functions of living, commerce, industry, government and social gathering in the face of calamities or hazards. In this case, urban resilience covers a number of systems. For resilience to be successful and sustainable, it should therefore be done in all the systems that make up the urban environment and give functionality to the urban areas. It is the capacity of urban systems, communities, individuals, organisations and businesses that make up the urban morphology to recover and maintain their functions after an economic or natural shock regardless of their impacts, frequency and magnitude. It is important to understand these definitions of urban resilience in the structuring and formulation of a resilience framework to know the main aspects that the framework is to tackle and the nature of stress in that the urban environment is to be relieved from. A resilience framework, therefore, provides a different urban aspiration and ideas about liveability and sustainability, cohesion development and robustness that consider chronic stressed and shocks the city is facing.

This paper seeks to develop a resilience framework that will be used to bring back into shape the urban areas of Zimbabwe that have been affected by different problems and to develop them to become liveable cities that are more sustainable and classic. A number of measures are to be set to build sustainability in as far as town planning is done and also to address issues, such as service delivery, living conditions of urban dwellers and economic growth of the urban areas that can be fostered by investment and the government's intervention in urban activities. Some of these measures include the devolution of power, creation of investments friendly environment and stabilisation of the currency, among others. The contribution of this resilience framework proposes the development of measures that contribute a lot in the outlook of the city and how the style of planning will be done. The economy will be improved and a huge number of investors will bring with them new technologies and innovation that will create employment and production at a faster rate. It can be urged that the reshaping of the urban areas of Zimbabwe and their development lies in the development of a very powerful and effective resilience framework that is capable of changing the way different activities are being done and the way in that the economy is being run. Therefore, for this framework to be effective in the transformation of urban areas of Zimbabwe, there is need to curb corruption and devolve power from central government, increase funding towards the renovation of the dilapidating urban areas and also find ways to raise the economic growth of the country as a whole. It is also important to uphold discipline and harmony to speed up the decision-making process. The study recommends that government should be the main actor in bringing about resilience and sustainability in urban areas.

Literature Review

The notion of resilience has become the topical discussion recently in the world. Though the idea has been there for a long time, it has gained popularity in the last decade. There are a number of reasons that has contributed to the application of resilience in the development of urban cities. One of the major calls for resilience is climate change and its effects on a global scale. Many countries have far adopted the concept of resilience to keep their countries and urban areas in a favourable living environment. At a global level, resilience has been done in most of the developed countries as a way of preserving their urban cities. A number of resilience frameworks have been formulated in different countries and organisations to address a number of issues. Some of these resilience frameworks include the climate change-based resilience framework, disaster management and how to overcome other stressful occurrences. Organisations, such as DFID, Oxfam and UN FAO also developed resilience frameworks that were meant to bring about sustainability and adaptation from different shocks and disasters in different countries and regions. The United Nations FAO (2014), developed a resilience framework in an attempt to bring together nutrition, livelihoods and risk reduction concerns under a resilience lens. These frameworks differ from one organisation to the other due to variations in programming, policy priorities, purpose and also the absence of a uniting definition of resilience. The resilience framework according to DFID (2011), has four elements that are context, disturbance, capacity and reaction. The context of the resilience framework is basically the social group, region or country that has been affected with a disaster or serious problem that needs to be addressed in a sustainable manner. The "disturbance" is the problem that has affected the proposed area in terms of insecurity, natural disasters or food shortages, among others. The "capacity" to deal with disturbances is another aspect that looks on the exposure, sensitivity and adaptive approaches to be implemented. The "reaction" to the disturbance is an element of resilience that focuses on the ability to survive, cope up with, recover and transform from a certain disaster or disturbance.

The world is faced with different problems that result in uncontrolled creation of non-liveable cities. These problems are socio-economic, political and environmental. There are certain measures that were put in place to reduce and curb adverse effects of these problems in urban cities. Different countries formulated structures and solutions that have the capacity to recover quickly or spring back into shape after deformities. For example, in China, urban spheres were faced with the problem of overpopulation. Myers et al. (2019) state that China has the highest population in the world, reaching 1.2 billion or 20 percent of the global population. This has led to challenges that include degradation of land and resources resulting in detrimental living conditions. The Chinese government has tried to find a solution to the problem of increased population. One of them is One Child Policy where the citizens will be offered benefits if they agreed to have a single child. This includes a tax of up to fifty percent for citizens who have more than one child (Brooks, 2018). Besides the one Child Policy, China used the sterilisation target goals that were set and made mandatory for people who had two children. This reduced the total population of China and raised the standards of living that even the number of people who could access taped water increased from 84 to 94% (Laidley, 2016). Africa's population is steeply increasing. The population of Lagos in Nigeria is expected to overtake Cairo as Africa's largest city (Bish, 2016). South America is noted to be one of the top continents that is urbanising in the world, with almost 84% of the total population living in the cities. Urban sprawl is another problem that has become a thorny issue in most countries nowadays. In the past decades, cities had defined and specificied boundaries but they have missed their realms by excessive growth today. It is noted that the total area covered by the worlds cities is set to triple in the next 40 years, eating up and threatening the planets sustainability.

Urban sprawl is referred to as the unrestricted outward expansion of the city. Urban sprawl results in the occupation of wetlands that will later cause water crisis as the earlier are lungs of the city (Maier & Winkel, 2017). Europe is one of the most urbanised continent on earth. More than a quarter of the continent's space has been taken over by urban land uses. This is caused by rural-urban migration where people move from rural areas to urban cities in search of greener pastures (Laidley, 2016). The city of Rotterdam; that has been strongly affected by urban sprawl, took a step-in densification that is a framework that leads to sustainability. The city introduced high-rise buildings that are not only a benefit in reduction of urban expansion, but also provide visual quality of the inner city. High-rise buildings accommodate more people at the same time saving more land. In China, most cities, like Beijing, introduced the method of building vertically to avoid urban sprawl (Zahar, 2017).

Air pollution is another major problem in the world cities. In countries like China, U.K, Japan, U.S.A, Italy, France and most of the European countries is rampant. The major cause is the use of modern technologies that heavily depend on fossil fuel, coal and oil. Pollution is the direct or indirect process by that any part of the environment is affected in such a way that it is made potentially or actually unhealthy, unsafe, impure or hazardous to the welfare of organisms that live in it (Mapira, 2015). Air pollutants accumulate in the atmosphere and cause global warming, depletion of the ozone layer and climate change. These will; in turn, cause temperature increase and result in natural disasters that affect the cities. Pollution is costly. It has resulted in welfare loss estimated at 4,6 million per year 6,2% of global economic output. Diseases caused by to pollution are responsible for 9 million premature deaths (Zahar, 2017). The combustion of fossil fuel and burning of biomass accounts for 85% of airborne diseases and also leads in depletion of the ozone layer. All these effects of pollution call for action.

In China, they introduced the Polluter Pays Principle in that anyone who disturbs or spoils the environment in any way has to take necessary corrective measures to rectify the environment or pay the cost of remediation. The framework also includes industries to absorb the negative externality they create. The Lancet Commission on Pollution and Health (2017) noted that in Malaysia, the city faced a problem of high land pollution and introduced Waste Recycling and Reuse. This framework reduced the rate of land pollution in the country. In the city of Rio de Janeiro Brazil and Kenya, congestion became a problem that it resulted in loss of economic welfare. The countries introduced the Local De- congestion Protocol that reduced the traffic congestion problem. In Johannesburg, South Africa, authorities introduced spaggeti roads more pronounced in highways from the OR Tambo airport. This enables the reduction of traffic congestion.

Shortage of clean, safe water is a global challenge that has been tackled differently by countries. Namibia is one country that has managed to promote sustainability in its water supply system. In an attempt to increase water supply that is one of the major problems in many arid and semi-arid regions, Namibia has engaged in water recycling and reuse in its urban cities. This has helped in the provision of clean water and eradication of water shortages. There have been water shortages due to inadequate water sources and this has led to the finding of alternatives; with the main one being the establishment of wastewater treatment plants, such as the New Goreangab Reclamation Plant in 2002, with the capacity of 21,000 m^3/day (Menge 2010). Windhoek city has also adopted the wastewater treatment method as a way of providing adequate water to consumers and also promote resilience and sustainability. Large effluents from the city are being treated and recycled into water for industrial, agricultural and other urban watering functions. It has been used for the creation of recreation sites, such as artificial lakes and ponds (Menge, 2010).

At national level, the problem of land pollution has proliferated in Zimbabwe. Inadequate provision of litter bins in urban areas and the inefficient collection of garbage in residential areas and the CBD is widespread. This has resulted in people throwing litter everywhere and polluting the environment. In some instances, in the vendors in the Central Business District use drains as dumping areas. This is blocking the water drains of the city and causing floods during rainy seasons. The idea of resilience is not new to Zimbabwe as there are steps that have been taken to promote it in both urban and rural areas around the country. Many communities that are funded by non-governmental organisations have introduced sustainable ways of increasing food security and uplifting of livelihoods in rural areas, such as Masvingo, Chipinge and Chiredzi, among others (Andersson, 2002). In urban areas, solutions have been introduced to solve problems, such as water shortages and pollution (Ndebele, 2009; Ndlovu, 2016). However, some of the measures that have been introduced in as far as resilience and sustainability is concerned were affected by a number of factors, such as corruption, lack of funding and poor governance. Some of the measures are short lived, therefore there is need for the development of a resilience framework that will be applied in urban cities of Zimbabwe to rebuild its economy and promote world standards that are liveable and well developed. The government introduced the clean-up campaign as a solution to the problem of land pollution. It declared every first Friday of the month as a national clean-up campaign day in all the cities in Zimbabwe. It can be viewed that the clean-up campaign is not sustainable as it is effectively done once a month living the other thirty days of the month polluted and undesirable. Due to lack of refuse collection, there is accumulation of garbage in residential areas and commercial centres (Tanyanyiwa, 2015).

The Environmental Management Agency (EMA) also put in place a method of reuse and recycle that is a strategy that is sustainable. Recycling is when things are converted to other usable material (Brown et al., 2018). It is important in reaching sustainability because it makes the materials that cannot decompose to be used for other purposes other than causing pollution. Also reusing is when a material is repaved and serves its initial function. This structure enables the reduction of land pollution in Zimbabwe. The Polluter Pays principle was also incorporated in Zimbabwe, where those who produce pollution should bear the costs of managing it to prevent damage to human health and environment. For example, companies, like Lafarge Cement that pollute the environment causing negative externality in society are committed to pay a tax that compensates for their damage. This tax can also be called the carbon pricing.

The Current Situation in Zimbabwe's Urban Areas

Urban areas in Zimbabwe are going through a phase of degradation. They are stressed with lots of problems that have gone out of hand due to a number of factors associated with the country's economy and governance. The problems that are affecting the urban areas of Zimbabwe can be grouped in a number of sectors. There are social, economic, environmental and political challenges that are stressing the urban areas.

The economic situation in Zimbabwe has reached the climax and is getting worse each day. Inflation is continuously rising while the monetary and fiscal policies that are being implemented are failing to solve the crisis. High unemployment rate is one of the major economic problems that are affecting the urban areas of Zimbabwe. It has contributed to the increase in street jobs, such as money-changing and vending. Harare and Gweru urban areas have been affected by the high figures of unemployment 700 firms in the country have closed down and many more companies are closing, throwing tens of thousands of workers into the streets to join the teeming ranks of unemployed that is estimated at above 88% (Hove et al., 2020). This has increased the number of street vendors and illegal activities in urban areas.

The increase in prices of basic commodities is another factor that is affecting the economic status of urban areas in Zimbabwe. Furthermore, lack of foreign investment has affected the economy of the cities. This has resulted in use of old-fashioned machinery in factories and some industries due to lack of new ideas and technology being introduced into the country by the investors. Not only do FDIs bring new technology, but also foreign currency that may be used to import some commodities that the country lacks in its urban areas to promote economic growth and development. Reduction in the value of the RTGS dollar has diverted the interest of investors. Thus, both foreign and local investors are not interested in making investments in the urban areas of Zimbabwe. Power shortage is another factor that has affected the economic growth of urban areas. The production rate of different companies has been forced to lapse as they are prolonged hours of darkness and incapacitation as the Zimbabwe Power Company (ZPC) is failing to produce enough electrical energy for the whole country. This has also reduced the percentage of exports. The number of goods that are imported is also increasing, thus taking out the little foreign currency that the country has that could be used to boost the economy of urban areas that are the centres of economic growth and development.

On the social side of urban areas in Zimbabwe, there is high rate of urban poverty and poor living conditions. The social systems are deteriorating, for instance, the health sector. The health facilities are operating poorly and a large number of people are losing their lives due to inavailability of medicine and also shortage of medical practitioners. In Harare, many health institution hospitals have been left operating with a small number of doctors, especially the main hospitals, including Parirenyatwa. Lack of staff for medical education and high drop-outs in the public health care posts resulted in many vacancies and inadequate supply of medical facilities in hospital and the death of many people (Madzimure, 2018). High crime rate and drug abuse have become more popular in urban areas. This is affecting youths, mostly in the high density surbubs where the rate of unemployment and poverty is very high. Drugs, such as or "mbanje", are being taken by a large number of youths. Commercial sex is also on its peak in cities, with some school children dropping out of school due to unavailability of school fees. They end up selling their bodies at a very cheap price to get money to buy basic needs for their families. This is also high in squatter settlements, such as Epworth, with a well-known place called Pabooster. Most avenues in Harare are packed up with young ladies selling sex for self sustainance at night due to the shortage of proper jobs. There is a high rate of rural to urban migration in search for greener pastures that has led to overpopulation within cities. The increasing population in urban areas has caused a number of problems, such as high pressure on the resources and services. Kamusoko et al. (2013) suggest that the population of Harare has been increasing at a faster rate since independence when migration controls were removed. The population has increased to 1,435,784 by 2012. Urban cities have been designed for a smaller population, hence the increase in number after independence has resulted in high demand on resources and services, such as water and sewer reticulation system that have not been expanded. This increase in population has caused an overload and destruction of these infrastructures. Shortage of housing is another problem that has hit urban areas of Zimbabwe and has contributed to a high number of informal settlements and squatter camps, such as Hopley slum in Harare. This is a typical example of urban sprawl.

The urban environment is also affected by high rate of pollution. Air, land, noise and water pollution have become the Achilles heel of urban cities. Increased volumes of traffic have been as a result of improved car ownership. Most of these vehicles are second hand cars and are being dumped into the country by Japan as they are no longer accepted to move in their country of origin due to pollution. This is resulting in higher emissions of carbon dioxide into the atmosphere, thereby causing global warming and climate change. There is also contamination of water sources by sewage systems that are bursting and have not been replaced since the time before independence. Besides the bursting of water sources, raw sewage is being diverted into water sources that are the major suppliers of water in urban areas. Lake Chivero has been the victim as raw sewage has been dumped into the dam. Most sewer discharge of Harare and Chitungwiza end up in Lake Chivero (Kangata, 2016).

Land pollution has become a major problem that is affecting the urban environment. Lots of landfills can be identified in most areas of the urban city. This is due to lack of garbage collection facilities and absence of adequate bins in the central business districts. Bulawayo as the second largest city in Zimbabwe has faced land pollution caused by illegal dumping of waste, poorly managed landfills and some waste disposal methods, emanating from commercial, industrial and households. This has caused health hazards to city dwellers (Ndlovu, 2016). Kamusoko et al. (2013) noted that services and amenities in high density suburbs where a huge number of people are located are very poor and inadequate. Poor service delivery has resulted in the absence of garbage collection in both the central business districts and residential areas too. Noise pollution has also increased due to the rising number of small and medium enterprises operating in the cities. Noise pollution is also a result of the increased number of vehicles, congestion and touting at bus terminus, making the urban area less conducive to live in.

The political situation in Zimbabwe has affected the development of these urban areas. The struggle for the control of urban councils among political parties has resulted in the devastating conditions of cities in recent years. This threatens the effectiveness and efficiency of service delivery. Service delivery in urban areas is determined by urban governance (Muchadenyika, 2014). According to him, urban governance is a sine qua non for the delivery of services in urban areas. This means that urban service delivery relies on the functionality of the governance system. Muchadenyika (2014), argues that political conflicts between the ruling party (ZANU-PF) and the opposition (MDC) has led to the diversion of resources meant for service delivery and development for political expediency resulting in poor services in urban areas. According to MDC (2013), the main reason for the collapse of urban areas is over-centralisation of power to local government ministers. There is a huge centralisation and politicisation of power within the urban governance.

A number of incidences of violence have also been noticed in urban areas due to the political conditions that surround city dwellers. It can be argued that there are authorities that suppress people from demonstrating against the hardship they are pressing on them, making their living standards poor. The political system has affected the functionality of urban areas, hence it is important to give an eye on it in the process of formulating a sustainable resilience framework for urban Zimbabwe. It can be urged that town planning in Zimbabwe is politicised. Most people that are involved in the planning of urban areas have political influence, thus making it difficult to bring about effective developments.

Most urban councils in Zimbabwe are run by the opposition party; that is the MDC-Alliance, while most of Members of Parliament and the local government are under the control of the ruling party (ZANU-PF). This delays the passing of judgements and affects the rate of development as there is no coherence between the two parties. Both of them use offices for political expediency. Government of Zimbabwe/ILO (2016) has noted political influence in the present situation in Zimbabwe. He observed that Zimbabwe's government cancelled debts owed by water users to the urban municipalities ahead of the 2013 national elections as in electioneering strategy. This act has caused a typical economic risk that has made water supply institutions to suffer. It has also affected and weakened the administration controlled by the opposition. Though the ruling party (ZANU-PF) managed to improve its popularity and votes in urban areas by disadvantaging the opposition party, the move reduced the resilience of urban water supply. In contemporary times, most suburbs have gone without municipality water coming out of their tapes, leading to the drilling of shallow. Wells that have resulted in diseases outbreak as the wells are contaminated by sewer from bursting pipes. Thus, the dire situation in the urban areas is also a result of the political grudges that exists between the opposition and the ruling party (Muchadenyika, 2015).

Towards a Resilience Framework for Sustainability

Resilience frameworks arise due to differing organisational programming, priorities and purposes and the fact that there is no agreed definition of resilience among international amongst members of the community. Resilience frameworks attempt to take a holistic and multi-sectorial approach that typically link humanitarian relief and disaster response to livelihoods. In most cases, these frameworks tend to address the sensitivity, exposure and adaptation to a shock or stress that has affected the community. Resilience frameworks focus on building capacities and also deals with disturbances. Therefore, this paper seeks to draw a framework that will be used to rectify the stress that is pressing the urban areas of Zimbabwe. As it is the role of a framework to bring about long lived solutions the study will address the way forward in improving the economic structure, social standards, political situation and the commercial status of urban cities.

In this chapter, it has been noted that there in need for formulation and implementation of a resilience framework in urban Zimbabwe. The problems that people in urban area are facing call for an action-oriented plan that is sustainable. They require a basic supporting measure or structure underlying a system with the ability and capacity to spring circumstances or facilities back into shape after some difficulties. Taking into consideration the problems of the urban areas of Zimbabwe, it was noted that the solutions should be able to address them for a long term and enable the city or affected area to recover back into their original state. Due to the discussed problems in different urban areas, such as Bulawayo, Gweru, Kwekwe and Harare, there is need to formulate a resilience framework that will help stabilise and restore the situation of urban areas back into shape, enhancing sustainability. Towards the minimisation of air pollution and congestion caused by the increased number of ex-Japanese cars that move in and out of the urban areas, like Harare, an embargo should be set to step their importation of vehicles and rely on the vehicles produced in Zimbabwe. This will also help increase the production and efficiency and uplifting the new and old car industry of Zimbabwe that has fallen apart due to the importation of cheap vehicles from Japan. There is need to improve on the public transport system in urban areas to promote a culture of commuting.

Strict laws need to be introduced towards pollution. High taxes showed be burdened on polluters and existing laws should be amended to increase penalties. In addition, law enforcement agents should increase vigilance on pedestrians who throw litter everywhere in the streets and cause them to pay lefty fines. Adequate supply of litter bins in Bulawayo and Harare CBD will also support the enforcement of the strict laws introduced. Since our urban areas are characterised by poverty and poor diet among the people mostly located in high density suburbs, there is need for the introduction of food aid in urban areas. Besides the food aid programmes, the idea of urban agriculture should be promoted to achieve food security. Land should be allocated for agricultural purposes to help feed the urban population. In an attempt to reduce poverty and raise the standards of living, employment must be created to allow the people to gain income that will enable them to run day to day living expenses. For employment to be created, there is need for investment. Both external and internal investors are required for the uplifting of the economy. Investors bring about new ideas, technology and employment creation for the jobless, hence government must create a conducive environment that will allow investors to be interested in investing in the urban areas of Zimbabwe.

Another key matter to be considered is the stabilisation of the currency. The Zimbabwean dollar is undergoing a high rate of inflation hence few people are ready to invest in such conditions. In this case, there is need to control the exchange rate and stabilise the currency so that people are attracted to invest in Zimbabwean cities. Another aspect that need to be addressed by this framework is urban design and settlement planning. Urban sprawl is at its peak, hence vertical planning of settlements should be done to promote densification in cities, like Harare, Bulawayo, Masvingo, Gweru and Kwekwe. High uprising buildings should be established for both commercial and residential purposes. This will create more space for other urban activities for the forth coming generations. Moreover, for urban areas to be developed and to be sustainable, there is need to separate planning and politics. The influence of politicians in urban activities has led to the current deplorable status of urban areas in Zimbabwe. Therefore, political power in urban development has to be reduced to in order allow transformation of the urban morphology. Urban areas should be planned in a professional way by people that have been trained to do so. Hence central government must relinquish power to city and town councils so that they tackle their role in ways that lead to sustainability and development at a faster rate. Thus, this paper adds voice to the call for devolution of power from government and the Ministry of Local Government.

The formulation and application of resilience framework should transform the status of the urban areas in Zimbabwe. It should contribute to a rapid development of urban areas and their further development in a sustainable manner. The urban infrastructure should reach another level with quality transport infrastructure, high standards buildings and involved water and sewer reticulation systems. The maintenance of water and sewer system should also reduce the outbreak of waterborne diseases that have been affecting urban areas, especially in Harare and Gweru. Adequate service delivery should also result in the creation of clean urban areas that are zero percent pollutant.

The provision of adequate bins and garbage collection around the shopping centre and the residential areas should be increased to erase the creation of landfills all over the places and the introduction of strict laws towards the dumping of rubbish along the streets will make the residents to be more disciplined towards the disposition of litter. This would improve the standards of living as people will enjoy a stable life style where they afford all the basics that are required for their day to day life. In the business sector, a huge number of investments should be identified to bring growth to the economy with new factories being opened and industries being upgraded through the introduction of new technology and different skills that increase the efficiency of the operating system. Furthermore, the urban environment should be transformed, making it the best place to reside in. There should be an increase with regards to urban green infrastructure through urban afforestation and increase in the number of public parks and golf courses. Besides that, the planting of land and other green places in residential areas, churches, institutions, school and offices will create a green environment that is welcoming for the growth of biodiversity. That will increase the value of the place and a relaxation environment. Moreover, agriculture contribute to the moderation of temperatures and carbon sequestration. Such resilience framework will also result in the exportation of products to the global market through investment that brings about new technologies and innovations that make the products to be demanded highly in different areas around the world.

The situation of the urban areas in Zimbabwe is worsening and action has to be taken to bring them back to shape. Many problems have been identified that are pressing these areas and how they have also affected the standards of living and the economic situation. There formulation of this resilience framework is very crucial in giving rise to improved urban standards that are expected at national and the global level. The sustainability and development of cities has been affected by a number of factors, such as corruption, poor governance and conflicts among political parties. Though the idea of resilience has been discussed before, a much further step needs to be taken to enhance sustainability in urban centres of Zimbabwe. There is need for the stabilisation of the economy, increase in investment, creation of jobs and green urban environments that are attractive. Infrastructure development is another key aspect that needs to be considered. Development should be monitored to balance the needs of the present generation and maximises the requirements of the generations to come, especially spatial distribution to create space for further development since land is a scarce resource. More opportunities have to be created to promote urban centres that are free from diseases and that are serviced, lacking nothing.

Conclusion

In the attempt to make urban cities of Zimbabwe more sustainable it should be noted that the resilience framework will help shape them and reach the ultimate goal of making them liveable places that the future generations will admire. The present residents need to acknowledge the resilience framework should be put into practice. As we have noted, many agendas have been set to guide the development goals of the country at a certain period but have failed due to a number of problems, such as corruption and lack of funding. Hence, if these mistakes are repeated, the framework will also remain a paper that ought to direct development without any change taking place on the ground. It is important that government and the local authority should start to act and give a good example towards the implementation of the framework to get the urban areas transformed into world first class cities. More funding needs to be put aside to sponsor the proposed developments that are set to take place in urban areas. It is also vital that the authorities should consider all the major steps that must be taken to speed development and to minimise the barriers that disturbs the follow of activities within the urban areas. It is recommended that in an attempt to promote resilience, government has to privatise some of its parastatals for them to produce more income and production. There is also need to dialogue with the private sector so that it may invest in new technology and innovations that will raise the country's market.

References

Andersson, A. (2002). The bright lights grow fainter: livelihoods, migration and a small town in Zimbabwe (Doctoral dissertation, Acta Universitatis Stockholmiensis).

Bish, J. (2016). Population growth in Africa: grasping the scale of the challenge. The Guardian, 11. Available online: https://www.theguardian.com/global-development-professionals-network/2016/jan/11/population-growth-in-africa-grasping-the-scale-of-the-challenge.

Brooks, A. (2018). *Africa, Development, Economy, Overpopulation.* The East Africa: Population growth.

Brown, C., Shaker, R. R., & Das, R. (2018). A review of approaches for monitoring and evaluation of urban climate resilience initiatives. *Environment, Development and Sustainability, 20*(1), 23–40.

Government of Zimbabwe/ILO. (2016). Employment Creation Potential Analysis by Sector. Available online: https://www.ilo.org/wcmsp5/groups/public/---africa/---ro-abidjan/---sro-harare/documents/publication/wcms_619735.pdf

Hamilton, W. A. H. (2009). Resilience and the city: The water sector. *Proceedings of the Institution of Civil Engineers-Urban Design and Planning, 162*(3), 109–121.

Hove, M., Ndawana, E., & Ndemera, W. S. (2020). Illegal street vending and national security in Harare, Zimbabwe. *Africa Review, 12*(1), 71–91.

Kamusoko, C., Gamba, J., & Murakami, H. (2013). Monitoring urban spatial growth in Harare Metropolitan province, Zimbabwe. *Advances in Remote Sensing, 2(4), 322–331.*

Kangata, A. (2016). *Act of Urbanisation on Water of Lake Chivero, Zimbabwe: Water and Environment*. Chinhoyi: Chinhoyi University of Technology.

Laidley, T. (2016). The problem of Urban Sprawl. *Contexts, 15*(3), 74–77.

Lancet Commission on Pollution and Health. (2017). *Pollution and Health: Six Problems & Six solutions.*

MAPIRA, J. (2015). Air pollution in Zimbabwe: An environmental health challenge. *Journal Of Social Science Research, 9*(1), 1750–1758.

MDC. (2013). *Election Manifesto 2013: A New Zimbabwe-The Time is Now*. Harare: Movement for Democratic Change, Information and Publicity Department.

Madzimure, E. (2018). *The effects of illegal gold mining by Globe and Pheonix mine retrenches on the environment*. Unpblished MSc Dissertation, Gweru: Midlands State University.

Maier, C., & Winkel, G. (2017). Implementing nature conservation through integrated forest management: A street-level bureaucracy perspective on the German public forest sector. *Forest Policy and Economics, 82,* 14–29.

Meerow, S., & Newell, J. P. (2015). Resilience and complexity: A bibliometric review and prospects for industrial ecology. *Journal of Industrial Ecology, 19*(2), 236–251.

Menge, J. (2010). Treatment of wastewater for re-use in the drinking water system of Windhoek. In *Water Institute of Southern Africa Conference: Midrand, Southern Africa.*

Muchadenyika, D. (2015). Land for housing: A political resource–reflections from Zimbabwe's urban areas. *Journal of Southern African Studies, 41*(6), 1219–1238.

Muchadenyika, D. (2014). *Contestation, Confusion and Change: Urban Governance and Service Delivery in Zimbabwe (2000–2012).* Unpublished PhD Thesis: University of Western Cape, South Africa.

Myers, S. L., Wu, J., & Fu, C. (2019). *China's Looming Crisis: A Shrinking Population.* New York Times, January 17, 2020. Available online: https://www.nytimes.com/interactive/2019/01/17/world/asia/china-population-crisis.html

Ndebele, M. R. (2009). Primary production and other limnological aspects of Cleveland Dam, Harare, Zimbabwe. *Lakes & Reservoirs: Research & Management, 14*(2), 151–161.

Ndlovu, F. (2016). *Land Pollution: Public Health Hazards and Risks in the City of Bulawayo. Faculty of Science and Technology*. Harare: Zimbabwe Open University.

Tanyanyiwa, V. I. (2015). Not in My Backyard (NIMBY)? The accumulation of solid waste in the avenues area, Harare, Zimbabwe. *International Journal of Innovative Research and Development, 4,* 122–128.

Zahar, A. (2017). *The Polluter Pays Principle and its Implications for Environmental Governance in China: Research Institute of Environmental Law*. Wuhan: Wuhan University.

Patience Mukuzunga is a B.Sc. Rural and Urban Planning student at University of Zimbabwe. She was born in Hurungwe and attended Primary school and secondary school at Seke 4 High School in Chitungwiza. She later proceeded to Advanced Level at Morgan High School in Harare.

Christine Chivandire was born at Gokomere Clinic in Masvingo on the 25th of December 1994. She went to Gokomere Primary School from 2000–2006 and joined Gokomere High School in 2007–2012. She is currently doing her Honours Degree in Rural and Urban Planning at the University of Zimbabwe.

Innocent Chirisa is a full professor at the Department of Demography Settlement & Development, University of Zimbabwe. He is currently the Dean of Faculty of Social and Behavioural Sciences the Faculty of at the University of Zimbabwe and a Research Fellow at the Department of Urban and Regional Planning, University of the Free State, South Africa. His research interests are systems dynamics in urban land, regional stewardship and resilience in human habitats.

Printed in the United States
by Baker & Taylor Publisher Services